Warm Dense Matter

Laboratory generation and diagnosis

IOP Series in Plasma Physics

Series Editors

Richard Dendy

Culham Centre for Fusion Energy and the University of Warwick, UK

Uwe Czarnetzki

Ruhr-University Bochum, Germany

About the series

The IOP Plasma Physics ebook series aims at comprehensive coverage of the physics and applications of natural and laboratory plasmas, across all temperature regimes. Books in the series range from graduate and upper-level undergraduate textbooks, research monographs and reviews.

The conceptual areas of plasma physics addressed in the series include:

- Equilibrium, stability and control
- Waves: fundamental properties, emission, and absorption
- Nonlinear phenomena and turbulence
- Transport theory and phenomenology
- Laser-plasma interactions
- Non-thermal and suprathermal particle populations
- Beams and non-neutral plasmas
- High energy density physics
- Plasma-solid interactions, dusty, complex and non-ideal plasmas
- Diagnostic measurements and techniques for data analysis

The fields of application include:

- Nuclear fusion through magnetic and inertial confinement
- Solar-terrestrial and astrophysical plasma environments and phenomena
- Advanced radiation sources
- Materials processing and functionalisation
- Propulsion, combustion and bulk materials management
- Interaction of plasma with living matter and liquids
- Biological, medical and environmental systems
- Low temperature plasmas, glow discharges and vacuum arcs
- Plasma chemistry and reaction mechanisms
- Plasma production by novel means

Warm Dense Matter

Laboratory generation and diagnosis

David Riley
School of Mathematics and Physics, Queen's University Belfast, Belfast, UK

IOP Publishing, Bristol, UK

ISBN 978-0-7503-2348-2 (ebook)
ISBN 978-0-7503-2346-8 (print)
ISBN 978-0-7503-2349-9 (myPrint)
ISBN 978-0-7503-2347-5 (mobi)

DOI 10.1088/978-0-7503-2348-2

Version: 20210401

IOP ebooks

British Library Cataloguing-in-Publication Data: A catalogue record for this book is available from the British Library.

Published by IOP Publishing, wholly owned by The Institute of Physics, London

IOP Publishing, Temple Circus, Temple Way, Bristol, BS1 6HG, UK

US Office: IOP Publishing, Inc., 190 North Independence Mall West, Suite 601, Philadelphia, PA 19106, USA

This book is dedicated to the memory of A Y Riley.

Contents

Preface

This book is intended as an introduction to experimental warm dense matter for graduate students entering the field. For that reason, I have aimed at presenting an overview that covers a broad range of topics rather than aiming for depth of detail in any one topic. Mainly due to the interests and experiences of the author, it is aimed at the experimentalist, and deals mostly with the methods of creating and diagnosing warm dense matter. Nevertheless, the first chapter gives an account of the important features of warm dense matter with reference to the main theoretical challenges. The second chapter deals principally with the use of shock waves in creating warm dense matter. Much of the warm dense matter related literature in this area deals with laser-driven shocks but we also cover ion-beams and gas guns, as well as explosive methods. In chapter 3, we cover the generation of warm dense matter by volumetric heating. This covers both the use of x-rays, usually from laser-plasmas, and the use of intense ion beams. In chapters 4 and 5, we discuss a variety of diagnostic techniques available in both the x-ray and optical regimes, respectively. Finally, chapter 6 gives an outline discussion of the large scale facilities that are most commonly used for warm dense matter research. There is much more detail to be found in the extensive literature and many more possible references than can be comfortably accommodated in one book. Thus, I have tried to select those that best illustrate the points under discussion whilst also trying to give due credit to earlier pioneering work.

In this field it is common to use units that are a mixture of SI and cgs systems and to use electronvolts to denote temperature as well as energy. I have not attempted to break away from this, but trust that the units used in any particular equation have been made clear.

Acknowledgements

Working in experimental physics is almost always a team activity, and I would like to express my gratitude to those many colleagues, post-doctoral researchers, and students with whom I have worked. Despite the often frustrating nature of carrying out warm dense matter experiments in a fixed, and often all too short experimental slot, it has mostly been a deeply rewarding experience, made so by the quality of people I have been lucky enough to work alongside. This book makes use of several figures from other authors, which are referenced and credited, but I would like to thank them all for agreeing to their use, here as well. I would also like to thank Gianluca Gregori, Steve Rose, Siegfried Glenzer and Jan Vorberger, who have all kindly given useful comments on parts of the manuscript.

Author biography

David Riley

Professor David Riley graduated with a BSc in physics from the University of Durham in 1984 and with a PhD from Imperial College, University of London in 1989. He has worked at Queen's University Belfast since 1994 and has authored or co-authored over 130 research papers in the general area of laser generated plasmas with many focussing on warm dense matter. He was elected a Fellow of the Institute of Physics in 2011.

Symbols

A	Atomic mass number
a	Ion-sphere radius
a_B	Bohr radius
$\mathbf{B}$	Magnetic field
c	Speed of light
D	Debye length
D_e	Electron Debye length
D_i	Ion Debye length
$\triangle U_i$	Ionisation potential depression
$(d\sigma/d\Omega)_T$	Classical Thomson scattering cross-section
e	Elementary charge
E_b	Electron binding energy
E_b	Compton scattering energy shift
$\varepsilon_{\mathrm{RPA}}(k, \omega)$	Plasma dielectric function in RPA approximation
ε_0	Permittivity of a vacuum
F	Degrees of freedom
f	Frequency (Hz)
f	Free-streaming limit in electron heat flow
$f(k)$	Ion form factor
$f_e(p)$	Fermi–Dirac electron distribution function
$G_{bf}(\omega, T_e)$	Bound-free Gaunt factor
$G_{ff}(\omega, T_e)$	Free-free Gaunt factor
γ_a	Adiabatic index (ratio of heat capacities)
γ_e	Adiabatic index for electrons
γ_i	Adiabatic index for ions
γ_G	Grüneisen parameter
γ	Lorentz factor
Γ	Strong coupling parameter
$\hbar$	Reduced Planck constant
h	Planck constant
I_A	Alfvén current limit
k_B	Boltzmann constant
k	Scattering wave-vector
κ	Thermal conductivity
λ	Wavelength
λ_β	Betatron wavelength
λ_{TF}	Thomas–Fermi screening length
Λ_e	Electron thermal de Broglie wavelength
μ_0	Permeability of a vacuum
μ	Chemical potential
$\mu(E)$	X-ray absorption coefficient
μm	microns (10^{-6} m)
n_c	Critical electron density
n_e	Free electron density
n_i	Ion density
n_r	Refractive index

ω	Frequency (radians s^{-1})
ω_p	Cold plasma frequency
ω_β	Betatron frequency
$q(k)$	Electron–ion correlation term
q	Reduced scattering wave-vector
Q_e	Electron heat flow
ρ	Mass density
ρ_c	Charge density
ρ_e	Resistivity
$\sigma(\omega)$	Electrical conductivity
σ_{SB}	Stefan–Boltzmann constant
$S_{ii}(k, \omega)$	Dynamic ion–ion structure factor
$S_{ii}(k)$	Static ion–ion structure factor
$S_{ei}(k)$	Static electron–ion structure factor
$S_{ee}(k, \omega)$	Dynamic electron structure factor
T_e	electron temperature
T_i	ion temperature
T_{hot}	Hot electron temperature in laser-plasma interaction
Θ_D	Debye temperature
u	Atomic mass unit
u_s	Shock speed
u_p	Particle velocity behind a shock
WDM	Warm dense matter
$\chi(E)$	Oscillatory part of absorption coefficient in XANES and EXAFS
$\chi(k, \omega)$	Density response function for a plasma
$\chi(p)$	Momentum space wave function for an electron
Z	Atomic number
$\bar{Z}$	Average ionisation
Z_b	Number of bound electrons
Z_p	Effective perturber charge
$Z(T, \rho)$	Partition function

IOP Publishing

Chapter 1

Background and context to warm dense matter

1.1 Introduction

Thanks to the efforts of many investigators over the last couple of decades [1–13], the experimental study of warm dense matter (WDM) is now a mature and well-developed branch of plasma physics. This state of affairs has come about, in large measure, due to the importance of WDM in the structure and formation of giant planets and other astrophysical objects, such as brown dwarfs, as well as in the implosion trajectories of proposed laser fusion capsules. Researchers are, of course, also motivated by the exciting technical challenges both in describing WDM theoretically and performing well defined laboratory experiments. In this book, the level of discussion is intended to be suitable for graduate students entering the field. Since the book is aimed at the experimentalist, we can leave a full and detailed discussion of the theory and modelling to the extensive literature already existing, some of which we shall reference along the way. Nevertheless, in this chapter, I will attempt to lay out the background and context of the subject, whilst pointing out some of the interesting aspects of WDM that are challenging from a theoretical viewpoint. The subsequent chapters will, by turns, discuss the principal methods for creating and diagnosing WDM.

Over the years, there have been many questions posed regarding the structure of the planets and the discovery of several thousand exo-solar planets in recent years [14–17] has only served to heighten our interest in the answers. For example, it has been proposed that carbon may form into diamond layers in planets such as Neptune and Uranus, due to the extreme conditions present [18]. In addition to this, new phases of matter have been predicted, for example, a superionic phase of water is predicted [19], where the protons are mobile within a lattice made from the oxygen atoms. This is predicted to occur at temperatures broadly in the range 2000–4000 K, for pressures of 30–300 GPa, and a similar state of matter has also been predicted for ammonia [20]. Recently, experimental x-ray diffraction measurements have found strong evidence of this phase of matter [21] for water.

doi:10.1088/978-0-7503-2348-2ch1

In addition to this, Nettelmann *et al* [22, 23] have demonstrated that, by utilising different equations of state (EOS) for hydrogen, alternative theoretical models of the interior of Jupiter can be created that have quite different internal structures and yet both of these models predict gravitational moments and other parameters, such as radius, and temperature close to the surface, that agree with data gathered in fly-by missions by Voyager, Juno, and Cassini [24]. This shows that observations of the planets themselves may not be enough to resolve theoretical questions regarding their structure and evolution, and experimental data taken in the laboratory is essential to our understanding of WDM.

1.2 Features of warm dense matter

It is natural that our first step in discussing WDM should be to give an approximate idea of what conditions we are speaking of, in other words; what do we mean by 'warm' and 'dense'? There are no precise and definitive answers here but, in figure 1.1, we show a map of the pressure–temperature space that covers the generally accepted

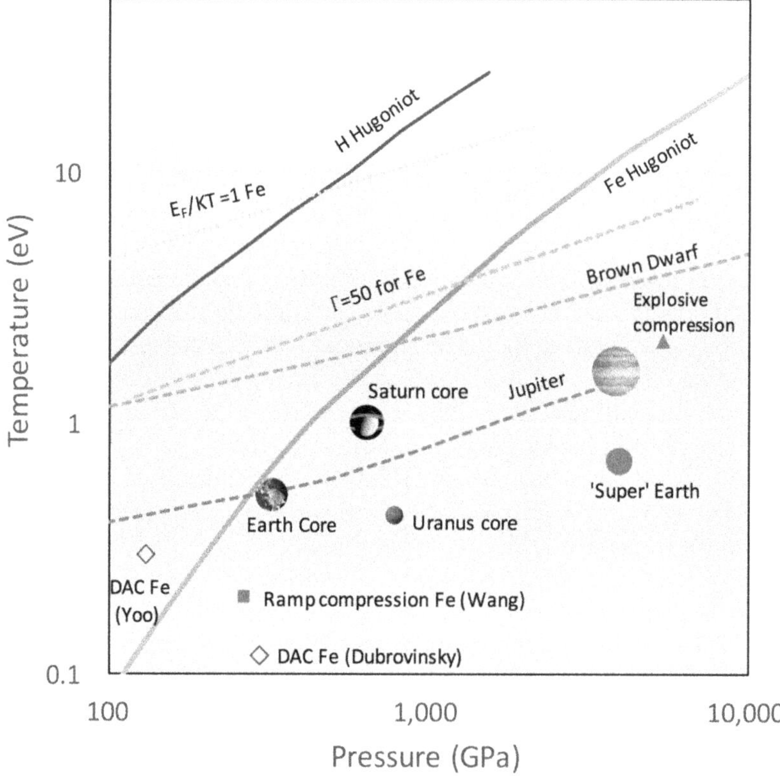

Figure 1.1. (Reproduced from [28].). Mapping of the warm dense matter region. The boundaries are a broad guide. We can see the way in which conditions in Jupiter as a function of depth (dashed blue line) pass through the WDM regime, as do conditions in a Brown Dwarf (dashed orange line). We note that static methods of compression and heating, such as diamond anvil cells (DAC), can only probe a peripheral part of the WDM regime.

regime for WDM. In broad terms, we are looking at states with pressures above a million atmospheres (100 GPa), and with temperatures between 1 and 100 eV (where 1 eV = 11 600 K). We have added the expected core conditions for some of the planets of our own solar system to the figure, as well as an indication of where we might see so called 'super-Earths' which are expected to have an iron dominated core at pressures and densities in excess of those existing inside our own Earth. An important thing to notice is that the conditions available via static compression using diamond anvil cells (DAC) only extend to the bottom left corner of our figure [25, 26], below the temperature range we generally consider to fall within the classification of WDM. As we see from the Hugoniot curves (see chapter 2), strong shock waves can generate a specific locus of states that pass through the regime of interest. As we will discuss in the next chapter, using ramp compressions [27] can give us access to states away from the Hugoniot, allowing us to explore the WDM regime more fully. It is usually with such strong shocks and ramp compressions that we can reach conditions appropriate to planetary interiors. Of course, not all WDM samples need to be at above solid density, and we shall discuss methods of WDM generation that do not require compression in chapter 3. However, as we will see below, our definition of WDM includes strong coupling between particles and this sets a broad lower limit of $\sim 10^{-3}$ g cm^{-3} to the range of densities of interest [1]. The fast timescales of WDM experiments, which can range from microseconds down to the sub-nanosecond domain, present experimental challenges that we will touch upon in this monograph.

WDM is not just defined by the range of densities and temperatures, but also by the physical phenomena that occur in this regime, and figure 1.1 also serves to remind us of some of the theoretical challenges that arise in describing WDM. These include strong coupling, degeneracy, and partial ionisation. We will discuss these separately, below, but it should become clear that they are interrelated issues.

1.2.1 Strong correlation

The first of the principal features of WDM that we discuss is strong inter-particle correlation, which can be expressed, for classical particles, through the parameter;

$$\Gamma = \frac{Q^2}{ak_BT} \tag{1.1}$$

where Q is the charge on a particle and a is a characteristic distance between particles. For ions, this is the ion-sphere radius, defined by;

$$a = \left(\frac{3}{4\pi n_i}\right)^{1/3} \tag{1.2}$$

where n_i is the average ion density. In the case of ion–ion correlations, this parameter has a value, $\Gamma \ll 1$, for low density and high temperature plasmas, such as are found in the solar corona or in a tokamak. What this means is that the potential energy of the Coulombic interaction between ions is small compared to the kinetic energy of their random thermal motion and can be treated perturbatively. Of course, it is a perturbation that leads to rich and complex physics. On the other hand, for solid matter, at temperatures below the melting point we have $\Gamma > 100$ and the thermal motion represents a small perturbation of an ion's position about a fixed lattice position. This situation can be dealt with by the use of approaches such as density-functional theory (DFT) [29]. For WDM however, we are generally in an intermediate range where neither the Coulombic nor thermal energy represents a small perturbation on the other. There is strong correlation between closely neighbouring particles but no long-range structure.

In figure 1.2, we can see an example of how the ion–ion correlation develops with strong coupling. We plot the radial distribution function, $g(r)$, for several values of Γ. This function is defined so that the number of ions found within a range of distances, r to $r + dr$, of a test ion is given by $4\pi r^2 dr g(r) n_i$. As we can see, as the coupling becomes weaker, there is little correlation and $g(r)$ tends to unity even for short distances. At stronger coupling, the repulsion between ions becomes more important and we see, at close distances, the development of a clear *correlation hole*, where other ions are repelled. However, repelling closer ions away from the test ion means that they are pushed towards farther ions which, in turn, push back. This leads to the oscillatory behaviour evident in the figure at intermediate distances. Moving to longer distances, the averaging is over a larger volume, and correlation with the test ion is weaker, thus the function averages out to unity.

The data in figure 1.2 has been calculated using a Monte-Carlo modelling approach [31, 32] that has assumed a so-called one-component-plasma. In this case, that means identically charged ions are immersed in a background of fully degenerate free electrons that are not polarised by the presence of the ions. In simple terms, the solution for $g(r)$ can be found by starting from some assumed, realistic, distribution of particles. Incremental changes to particle positions can be made randomly, to generate many configurations that will each have an energy, E, determined by the assumed inter-ionic potential. We can assume that the number of times configurations of a given energy can occur is proportional to the Boltzmann factor, $\exp(-E/k_B T)$, and this can be used to accept or reject new configurations, thus building up our distribution. The assumption of a fully degenerate electron gas,

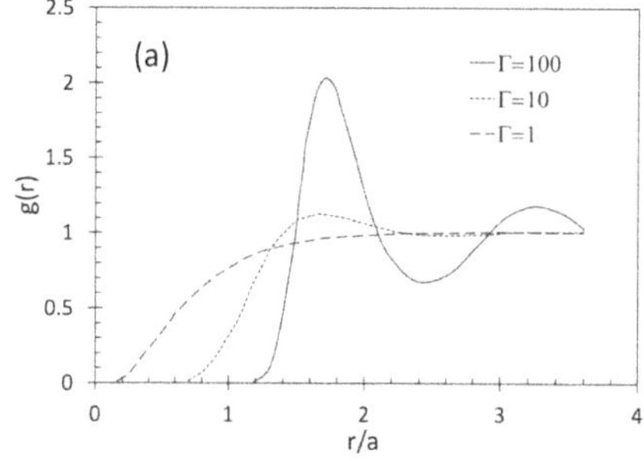

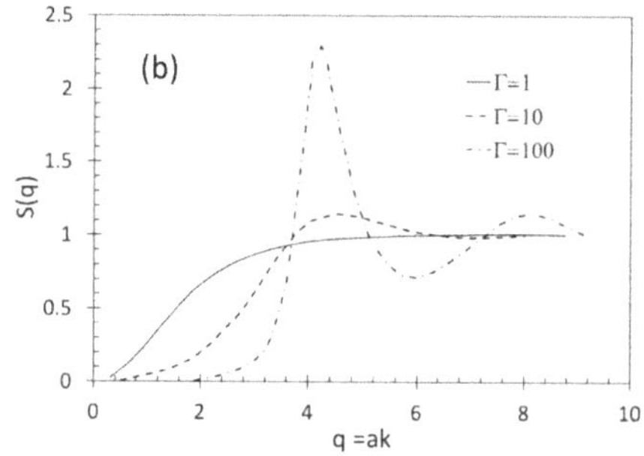

Figure 1.2. (a) The ion–ion radial distribution function for plasmas with varying degrees of strong coupling. The data is taken from [30]. The spatial scale is scaled to the ion-sphere radius, a, given by equation (1.2). (b) Structure factor for a one-component plasma model derived from the data in (a) and using equation (1.9). The parameter q is the dimensionless reduced wave-vector.

used in figure 1.2, is only valid in extreme conditions. In reality, degeneracy will be partial to some degree and polarisation of the electrons means that they can gather preferentially around the ions and cause screening. In a low density plasma, this is usually represented by a Yukawa type potential between ions;

$$V(r) = \frac{(\bar{Z}e)^2}{r} e^{-r/\lambda_{sc}} \tag{1.3}$$

where λ_{sc} is an effective screening-length and $\bar{Z}e$ is the average charge on an ion. For plasmas where $\Gamma \ll 1$, the Debye–Hückel screening length is appropriate and, for electrons, this is given by;

$$D_e = \sqrt{\frac{\varepsilon_0 k_B T_e}{e^2 n_e}} \tag{1.4}$$

The Debye–Hückel screening model is predicated on there being a large number of particles within a *Debye sphere*. This *Debye number*, N_D, is related to the strong coupling parameter via;

$$N_D = \frac{1}{(3\Gamma)^{3/2}} \tag{1.5}$$

where we see that, for strongly coupled plasmas, N_D is below unity and the Debye–Hückel approach should no longer be valid. For WDM cases, it is common to replace the Debye length with a finite temperature Thomas–Fermi screening length [33] defined by;

$$\frac{1}{\lambda_{TF}^2} = \frac{4e^2 m_e}{\pi \hbar^3} \int f_e(p) dp \tag{1.6}$$

in which we have made use of the Fermi–Dirac momentum distribution, $f_e(p)$, for a finite temperature. In the discussion of degeneracy below, we will see that the number of particles in a sphere of radius λ_{TF} is still less than unity for a wide range of WDM densities and that, in fact, more sophisticated approaches to screening are really needed. This is an important technical challenge for modelling of WDM. Nevertheless, as shown, for example, by Vorberger and Gericke [34], it can be reasonably accurate, for a range of WDM relevant conditions, to model the inter-ionic potential with Yukawa type, using the Thomas–Fermi screening length, and this is often done.

A commonly implemented alternative to Monte-Carlo methods and DFT, that is a relatively fast method to obtain the ion–ion structure factor, is to take the Ornstein–Zernike equation from liquid state theory [35, 36]. This uses a prescribed inter-ionic potential, $V(r)$, and relates a *direct* correlation function, $c(r)$, which accounts for interaction directly between two ions to a *total* correlation function, defined by $h(r) = g(r) - 1$, which accounts for correlation between ions via their effect on a third ion;

$$h(r) = c(r) + n_i \int c(r) h(|r - r'|) dr' \tag{1.7}$$

This is generally used with the hypernetted-chain (HNC) approximation which relates these functions to the potential;

$$-\frac{V(r)}{k_B T} = \ln(h(r) + 1) - (h(r) - c(r)) \tag{1.8}$$

These equations are solved numerically, typically by making an initial guess at $g(r)$ using $c(r) = h(r)$ in equation (1.8). A new value for $h(r)$ is then obtained from equation (1.7) and iteration to a self-consistent solution proceeds. In practice, it is usual to use the Fourier transforms of the functions for the iteration as this takes advantage of the ability to write equation (1.7) as $\tilde{h}(k) = \tilde{c}(k) + n_i\tilde{c}(k)\tilde{h}(k)$, where the wave-vectors, k, represent the spectrum of density fluctuations in the sample. We shall see an example of the use of the HNC approximation in figure 1.4 below.

The microscopic arrangement of the ions, created as a result of strong coupling, has a significant effect on such macroscopic quantities as the electrical resistivity, compressibility, and internal energy of a sample. This is often expressed through the static ion–ion structure factor, $S_{ii}(q)$, defined by a Fourier transform of $g(r)$;

$$S_{ii}(q) = 1 + 3\int_0^\infty \frac{\sin(qr)}{qr}[g(r) - 1]r^2 dr \tag{1.9}$$

where $q = ak$ is a dimensionless, reduced, scattering wave-vector. In figure 1.2(b), we can see the structure factors derived from the radial correlation functions in figure 1.2(a). As an example of the importance of structure factors, we can see how they are central to the Ziman form of dc electrical resistivity [37, 38] in a highly degenerate plasma;

$$\rho_e = \frac{m_e^2}{12\pi^3\hbar^2 e^2 n_e}\int_0^{2q_F} q^3|V_{ei}|^2 S_{ii}(q)dq \tag{1.10}$$

where q_F is the electron momentum at the Fermi level and we have an electron–ion potential, $V_{ei}(q)$. In addition, it can be shown that the compressibility at constant temperature of a sample is related to its structure factor at zero reduced vector;

$$S_{ii}(0) = \frac{n_i kT}{\rho(\partial P/\partial\rho)_T} \tag{1.11}$$

where n_i is the ion density. We can see from figure 1.2(b) that, for a OCP, $S(q)$ tends to zero at low q and thus is predicted to be incompressible, which arises as a result of the electron degeneracy. In reality, of course, the structure factor does not go to zero and a great deal of effort goes into theoretical modelling of the correct behaviour, which depends on the inter-ionic potential. This ion–ion structure factor will also play a key role in the discussion of x-ray scattering in chapter 4, where we shall present some experimental data that is compared to models more realistic than the OCP model.

1.2.2 Electron degeneracy

Another quantity that features in figure 1.1 is the ratio of the chemical potential, μ, for the free electron gas to the electron thermal kinetic energy, represented by k_BT_e. This tells us the degree to which we have partial free-electron degeneracy. For finite temperature, the chemical potential is given by;

$$n_e = 2\frac{(m_ec^2k_BT_e)^{3/2}}{\sqrt{2}\pi^2(\hbar c)^3}F_{1/2}\left(\frac{\mu}{k_BT_e}\right) \tag{1.12}$$

where n_e is the free electron density and $F_{1/2}$ is the complete Fermi–Dirac integral;

$$F_{1/2}(\eta) = \int_0^\infty \frac{x^{1/2}}{1 + \exp(x - \eta)}dx \tag{1.13}$$

with $\eta = \mu/k_BT_e$. This function has been tabulated [39] and the tables have been fitted in easily codeable analytical forms for different regimes of η, for example by Latter [40].

At the low temperature limit, the chemical potential reduces to the Fermi energy given by;

$$E_F = \frac{\hbar^2}{2m_e}(3\pi^2n_e)^{2/3} \tag{1.14}$$

For typical solid or liquid conditions, we can easily find $n_e > 10^{23}$ cm^{-3}, and so get $E_F > 1.25 \times 10^{-18}$ J (7.8 eV). The Fermi temperature is defined by $T_F = E_F/k_B$ where k_B is the Boltzmann constant. For known materials, the melting point under ambient conditions is only a fraction of an eV and so for liquid metals, for example, we can approximate that the electrons are in the ground state. This means that the chemical potential and Fermi energy are very similar. In fact, for the high degeneracy limit, the Sommerfeld expansion can be used to relate the two;

$$\mu = E_F\left[1 - \frac{\pi^2}{12}\left(\frac{k_BT_e}{E_F}\right)^2\right] \tag{1.15}$$

which is accurate to around 0.5% up to $k_BT_e/E_F \sim 0.3$ and higher order expansions can be used [41]. However, for WDM, where temperatures can be of order 10 eV or more, we can easily find that $\eta \sim 1$, and we can no longer assume that the electrons are in the lowest possible energy states.

Another way that we can express the degree to which degeneracy is important, under some particular set of conditions, is to compare the average distance between electrons with their typical spatial extent, as given by the thermal de Broglie wavelength;

$$\Lambda_e = \sqrt{\frac{2\pi\hbar^2}{m_ek_BT_e}} \tag{1.16}$$

We can check for the overlap of electron wavefunctions by looking for the condition;

$$n_e \Lambda_e^3 > 1 \tag{1.17}$$

which indicates that multiple electrons are occupying the same physical volume and degeneracy is important. For a 10 eV sample, this condition is fulfilled at $n_e \sim 10^{23}$ cm^{-3}, whilst, for the ions, the equivalent expression remains $\ll 1$. For typical WDM experiments, we are always dealing with classical ions. This partial degeneracy is important because the Pauli exclusion principle plays a role in determining such properties as the electron–ion equilibration time and associated quantities, such as plasma resistivity. Such processes are governed by collisions that change the energy of free electrons and our equations for these properties need to be modified. As an example, degeneracy will modify the electron-thermal conductivity [42–44]. Compared to the classical Spitzer [45] conductivity, there is a correction factor, ϕ_{cond}, which can be given as;

$$\phi_{\text{cond}} = \frac{\sqrt{\pi}\left[15I_2(\eta)I_4(\eta) - 16I_3^2(\eta)\right]}{144I_{1/2}(\eta)I_2(\eta)} \tag{1.18}$$

where

$$I_n(x) = \int_0^\infty \frac{y^n dy}{e^{y-x} + 1} \tag{1.19}$$

As we can see from figure 1.3, degeneracy tends to increase electron thermal conductivity as lower energy states into which thermal electrons may be scattered are blocked by Pauli exclusion, thus reducing the rate at which the electrons can lose energy to scattering. We shall see, in the following chapters, that with the development of short pulse optical and x-ray lasers, there are experiments that have the potential to allow determination of the electron–ion equilibration times.

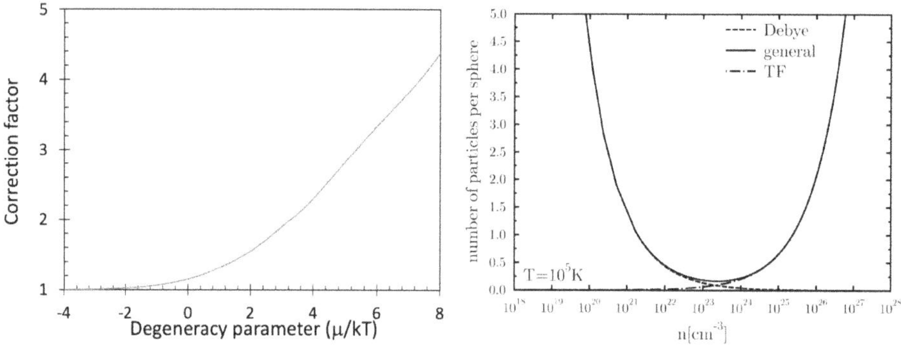

Figure 1.3. (a) Correction to the classical electron-thermal conductivity as a function of degeneracy, as given by equation (1.18) [42]. The conductivity increases with degeneracy because final states for electrons to be scattered into become Pauli blocked. (b) Number of particles in a Debye sphere, a fully degenerate Thomas–Fermi sphere, and a finite temperature model [46] that spans all degeneracies, as in equation (1.21), as a function of electron density for a temperature of 10^5 K. Calculation and figure courtesy of J. Vorberger.

Degeneracy will also alter some other key plasma parameters. As discussed in section 1.2.1, one of these is the Debye screening length, which, for electrons in a classical plasma, is given by equation (1.4). This is generally replaced by the Thomas–Fermi screening length, which, in the fully degenerate electron gas limit, is given by;

$$\lambda_{TF} = \sqrt{\frac{2\varepsilon_0 E_F}{3e^2 n_e}} \tag{1.20}$$

For finite temperature, with partial degeneracy, this can be replaced by equation (1.6), or can alternatively be expressed in terms of the Fermi–Dirac integrals;

$$\frac{1}{\lambda_{TF}^2} = \frac{n_i e^2}{\varepsilon_0 k_B T_e} \frac{F_{1/2}^{'}(\eta)}{F_{1/2}(\eta)} \tag{1.21}$$

where, as above, $F_{1/2}(\eta)$ is the Fermi–Dirac integral and $F_{1/2}^{'}(\eta)$ is its first derivative. This expression yields the Debye length for low density, non-degenerate plasmas and the Thomas–Fermi length at the fully degenerate limit.

In the previous section, we discussed the strong correlation that can exist between the ions, especially at high density and modest temperature. A consequence of the degeneracy of the electrons is that, in contrast to the ions, the coupling parameter for electrons gets smaller at high density. In order to define an electron coupling parameter for fully degenerate electrons, we can use the Fermi energy in equation (1.1), instead of the temperature, and obtain;

$$\Gamma_{ee} = \frac{e^2}{a_e E_F} = 0.543 r_s \tag{1.22}$$

where, for the electrons, a_e is defined as for the ion-sphere radius in equation (1.2), and the dimensionless electron-sphere radius, r_s, is defined by;

$$r_s = \left(\frac{3}{4\pi n_e}\right) / a_B \tag{1.23}$$

where a_B is the Bohr radius. We see that, when the density rises, the coupling will get smaller as $\sim 1/n_e^{1/3}$. As we can see in figure 1.3(b), with rising density, the number of particles in a sphere of radius given by the Thomas–Fermi length rises, whilst the coupling gets weaker. This perhaps goes part way to helping explain the relative success of screened Yukawa type potentials in WDM simulations.

For working between regimes, where we have partial degeneracy, it is worth noting that Gregori [47] has presented a treatment that interpolates between the two extremes, defining an effective quantum temperature in terms of the Fermi temperature;

$$T_{\rm qm} = \frac{T_F}{(1.3251 - 0.1779\sqrt{r_s})} \tag{1.24}$$

An effective temperature for the intermediate regime is then constructed as;

$$T_{\text{eff}} = \sqrt{T_{\text{qm}}^2 + T_e^2} \tag{1.25}$$

This effective temperature is then used in standard plasma physics formulae to derive the value of parameters such as screening lengths for the intermediate regime.

1.2.3 Partial ionisation

In addition to degeneracy and strong coupling, what may not be so evident from figure 1.1 is that partial ionisation is also a key factor in the behaviour of WDM, especially for mid to high-Z elements. The presence of bound shells of electrons can have a significant effect on the inter-ionic potential between ions. This arises at high density as the bound shells of neighbouring ions get closer to each other and a short-range repulsive term becomes important [48].

This naturally affects the microscopic arrangement of ions within the sample, as we can see in figure 1.4. In this figure, we show three simulations for an Fe plasma at $\rho = 14.1$ g cm^{-3} and $T = 3$ eV with $\bar{Z} \sim 4.9$ determined from the Thomas–Fermi model. One curve is for the OCP model where we assume the ions move under the influence of a bare Coulomb potential and the strong coupling parameter, $\Gamma \sim 100$.

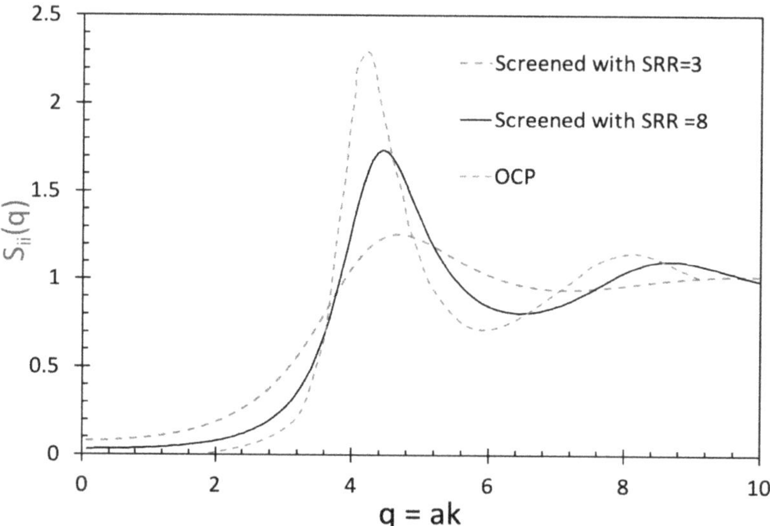

Figure 1.4. In this figure we compare an OCP calculation of the ion–ion structure factor to HNC calculations made for an Fe plasma at 14.1 g cm^{-3} and $T_e = 3$ eV with $\Gamma \sim 100$. For the HNC cases, a screened potential is used with a short-range repulsive term to simulate the bound-shell interactions. As we can see, even for this dense and highly degenerate sample, the OCP model is only a rough approximation. We also note that the relative strength assumed for the short-range repulsive term can have a significant impact on the shape of the structure factor and a small effect on the position of its peak. The HNC code used is the THEMIS code [33], access courtesy of Dirk Gericke.

In the other two curves, we have used a hyper-netted chain approximation to model the same plasma but we have used a potential of the form;

$$V(r) = \frac{SRR}{r^4} + \frac{(Ze)^2}{r} e^{-r/\lambda_{TF}} \tag{1.26}$$

where the second term is a Yukawa type potential, with a Thomas–Fermi screening length, and the first, short range, term represents the effect of repulsion between the shells of bound electrons with some empirically set constant, *SRR*, that can be varied. We can see from figure 1.4 that both the polarisation of electrons to allow for screening of the ions from each other and the strength of the short-range repulsion between bound orbitals, that are being forced together, can have a profound effect on the ion–ion spatial correlation.

As a summary of this section, we can look to figure 1.5, where we have used a Thomas–Fermi model to calculate the fractional ionisation for magnesium ($Z = 12$) over a wide range of conditions and then used this to derive the strong coupling parameter and electron degeneracy. We will discuss the physics of the Thomas–Fermi model below. The method of numerical solution used is outlined in Latter [40] and is without exchange or gradient effects included. The boundaries of the shaded region should not be taken as too definitive, but they do show the regime where we can say that we have partial degeneracy, i.e. $10 > \eta > 0.1$, and that we also have

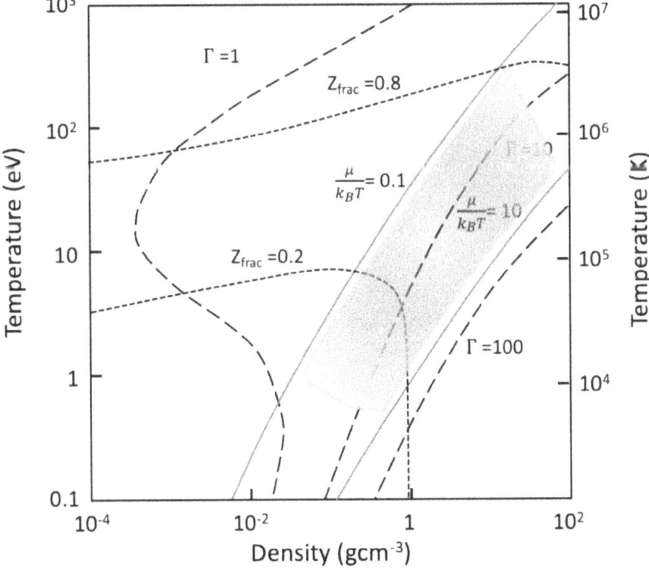

Figure 1.5. In this figure, we have used the Thomas–Fermi model to calculate the fractional ionisation (short-dashed line), ion–ion strong coupling parameter (long-dashed line) and the chemical potential divided by temperature (solid red line) for a Mg plasma over a wide range of densities and temperatures. The Thomas–Fermi model is likely to be more applicable in some regimes than others but the figure shows a broad guide to the region where the three characteristics of warm dense matter discussed in this section co-exist. In this figure $Z_{\rm frac} = \bar{Z}/Z$, where $\bar{Z}$ is the average ionisation and Z is the atomic number.

strong coupling with $\Gamma > 1$. At the same time, we see that the ions are not fully ionised and bound electron shells will be present.

1.3 Effects of warm dense matter on electronic structure

An important feature of dense plasmas and WDM is the plasma micro-field surrounding the ions and atoms. Fluctuations in this micro-field lead to Stark broadening, a subject that has been extensively explored for many decades [49]. As we shall see, emission spectroscopy, where line-broadening is an important diagnostic, is not common in WDM studies because the high density and low relative temperature preclude emission from the bulk of the sample. However, Stark broadening at high density can cause bound levels to merge with one another and the continuum, thus affecting the absorption spectrum. The surrounding micro-field also has a significant effect in producing continuum lowering and pressure ionisation, which are important in determining the electronic properties of the ions and thus affect the bulk properties of the sample. For example, an effect of WDM creation may be the metallisation of samples that are not metals initially. This can be a consequence of pressure ionisation and changes to band structure, as a sample is heated and/or compressed. Indeed, the question of whether, in a planetary core, hydrogen under goes a first order transition to a metallic fluid [50–52] with a discontinuity in density and entropy, or a continuous transition, is a subject of great importance for planetary evolution.

1.3.1 Continuum lowering

As we have noted above, in WDM, we have ionisation of the constituent elements and the ions created sit in a sea of free electrons. One of the principal effects of this is the phenomenon of continuum lowering or ionisation potential depression. In this, the potential generated by the surrounding plasma lifts the bound energy levels, lowering the ionisation potential. For higher lying levels, this can result in states becoming unbound and if these are occupied states, this causes pressure ionisation. It is generally recognised that there are two limiting cases, one at low density and one at high density. In the low-density case, for a given ion stage, Z_i, the continuum lowering, ΔU_i is given by the Debye–Hückel model;

$$\Delta U_i = \frac{Z_i e^2}{4\pi\varepsilon_0 D} \tag{1.27}$$

where D is the Debye screening length for the plasma. In the low-density limit, for non-degenerate electrons, this can be given by;

$$\frac{1}{D^2} = \frac{1}{D_i^2} + \frac{1}{D_e^2} = \frac{Z_p n_e e^2}{\varepsilon_0 k_B T_i} + \frac{n_i e^2}{\varepsilon_0 k_B T_e} \tag{1.28}$$

where Z_p is the effective perturber charge of the surrounding plasma, given by $Z_p = \langle Z^2\rangle / \langle Z\rangle$, and is typically similar to the average ionisation, and the other symbols have their usual meaning. We are more interested in cases at the other extreme of high density, where we have strong coupling and partially degenerate electrons. In these cases, we can replace the electron Debye length using the expression in equation (1.21). However, as we have noted already in figure 1.3(b), the number of particles in a sphere of radius given by the screening length, remains low for a wide range of WDM conditions and, at high density, it is common to consider a different approach with the ion-sphere approximation. In this limit, the ion-sphere radius is defined as;

$$R_i = \left(\frac{3Z_i}{4\pi n_e}\right)^{1/3} \tag{1.29}$$

and for the case where, due to the high density, we have equal ion and electron temperatures, the continuum lowering is given by;

$$\Delta U_i = -\frac{3}{2}\frac{(\bar{Z}+1)}{\bar{Z}}\frac{Z_i e^2}{4\pi\varepsilon_0 R_i} \tag{1.30}$$

where $\bar{Z}$ is the average ionisation of the perturbing ions surrounding the test ion in question, Z_i. In between these limits, the treatment of Stewart and Pyatt [53] has been the standard for plasma physics since the 1960s. More recently, as a result of the work of Ciricosta *et al* [54] and Hoarty *et al* [55], there has been some discussion of the correct treatment of continuum lowering [56, 57]. In the former case, observations of emission from a solid density Al target, irradiated with an intense x-ray beam from the LCLS free-electron x-ray laser facility, led to the conclusion that an earlier model by Ecker and Kröll [58] fitted the data better. However, the experiments of Hoarty *et al* on shock compressed Al, heated with fast electrons, led to the conclusion that the Stewart–Pyatt model was preferred. These apparently conflicting conclusions have led to a resurgence in interest in this subject after some decades of generalised acceptance of the Stewart–Pyatt model.

The magnitude of the continuum lowering effect can be substantial. For example, if we consider the case of warm dense Al at solid density and a temperature of 10 eV, we can calculate the continuum lowering in the ion-sphere approximation as ~ -45 eV for $Z_i = 3$. This is a substantial effect and energy levels such as the $n = 3$ are expected to be unbound, as with the ambient solid. As we shall discuss in chapter 4, one way of investigating this phenomenon is to look at the changes in the K-edge for WDM. However, this is not straightforward as, on heating the sample, we also change the degeneracy of the free electrons and, usually more significantly, the average ionisation, which means the shifts are a net result of multiple effects and are in fact smaller by an order of magnitude than the ionisation potential depression.

It has been further pointed out by, for example, Iglesias and Sterne [59] that the ionisation potential depression will not be exactly the same for all ions at the same time in a dense plasma, there will be fluctuations which complicate the direct study of IPD. They develop a model to deal with the fluctuations in which the number of electrons in an ion-sphere is calculated as a Poisson distribution around a mean value given by the net charge on the ion in question. These fluctuations are related to the fact that the curves of the ion–ion correlation function in figure 1.2 are a measure of the probability of the position of the nearest neighbour being at a particular distance. This means that the local electric field experienced by a 'test ion' will also show a distribution. This can be compared to the situation in plasma spectroscopy, where the plasma microfield can manifest itself in quasi-static Stark broadening of emission lines [49] whose shape reflects the micro-field.

1.4 Equations of state for warm dense matter

In the sections above, we have considered the characteristic features and some of the physics relating to WDM. Ultimately, a key goal is to construct realistic EOS that connect the bulk properties of density, temperature and pressure. In modelling experiments, it is of course desirable to have EOS that can be rapidly calculated for the wide range of time and space dependent conditions that can occur in a sample. There has been a great deal of work in this direction. In some cases, tabular EOS are built up from a combination of experimental data and theoretical approaches, with the latter differing according to the range of conditions being considered. An example of this is the SESAME equation of state library [60]. This data base contains a wide selection of elements as well as numerous commonly used non-elemental materials such as polystyrene and glass.

There are some key advantages of tabular EOS. For example, they are relatively easy to implement in a radiation hydrodynamics simulation code. The pressure in any one simulation cell at any time-step can be obtained by interpolation of the tables. Secondly, where it is available, experimental data can be used to constrain the tables and where there is no data, different theoretical models appropriate to different regimes can be applied to construct the table. On the other hand More [61] has discussed some potential drawbacks to tabular EOS. Amongst others, one problem is the issue of coarse spacing in tables that can lead to interpolation errors and inaccurate values of derivatives needed, for example for calculations of sound speeds. Another problem may be that a particular material is not tabulated. In the following, we shall discuss some of the physics that can go into constructing an equation of state.

1.4.1 The Thomas–Fermi model

Perhaps the simplest equation of state suitable for the WDM regime is the Thomas–Fermi (TF) model. This model was originally used to describe isolated atoms but has been adapted for use in high density matter under extreme conditions [40, 62]. A detailed description of its use in high energy density matter is given by Salzmann [63] but we will present a brief overview here. The basis of this model is to couple Fermi–Dirac statistics for the electron energy distribution with the Poisson equation for the potential around an atomic nucleus, where for the electrons we have;

$$\nabla^2 V_e(r) = \frac{eN_e(r)}{\varepsilon_0} \tag{1.31}$$

and for the nuclei we simply have;

$$V_{\mathrm{nuc}}(r) = \frac{Ze}{4\pi\varepsilon_{0r}} \tag{1.32}$$

Unlike application of the Thomas–Fermi model to the isolated atom, however, the electrons are constrained to lie within the ion-sphere radius, so that electrical neutrality is maintained within the ion-sphere. We should also note that spherical symmetry is assumed. Fermi–Dirac statistics gives us the electron energy distribution in the presence of the potential as;

$$n_e(r) = \frac{8\pi}{h^3}\int_0^\infty \frac{p^2\,dp}{\exp[(p^2/2m_e - eV)/k_BT + \eta]} \tag{1.33}$$

where p is the electron momentum and we assume $T = T_e = T_i$. This gives the result;

$$n_e(r) = 2\frac{(m_e c^2 k_B T_e)^{3/2}}{\sqrt{2}\,\pi^2(\hbar c)^3} F_{1/2}\left(\frac{\mu + eV(r)}{k_B T}\right) \tag{1.34}$$

Note the similarity to equation (1.12), except that now there is a position dependence and the potential is included. The equations above are the core of the Thomas–Fermi theory and are generally solved by numerical iteration [40] to get the radial electron density and potential self-consistently.

A well-known property of the Thomas–Fermi model is the way in which the solutions scale with Z so that solutions look the same for any material, as long we scale the temperature and density appropriately. As discussed, in detail, by More [4], if we take a given atomic number Z and atomic mass, A, at density ρ and temperature T, we can define a scaled density and temperature;

$$\rho_0 = \frac{\rho}{AZ} \tag{1.35}$$

and

$$T_0 = \frac{T}{Z^{4/3}} \tag{1.36}$$

Then we will generate a scaled pressure given by;

$$P_0 = \frac{P}{Z^{10/3}} \tag{1.37}$$

By taking the values of ρ_0, T_0 and P_0, we can find new values of these for another Z and A by simply using these scalings.

The bound and free electrons are treated on an equal footing but we can calculate the ionisation in a couple of ways. The first, commonly used method, is to ignore polarisation of the free electrons by the attraction of the nucleus and take the electron density at the ion-sphere boundary, with the convention that $V(r) = 0$ at this position. This means that the electrons at the boundary are free and we assume they are uniformly distributed throughout the ion sphere.

$$\bar{Z} = \frac{4\pi}{3} R_0^3 n_e(R_0) \tag{1.38}$$

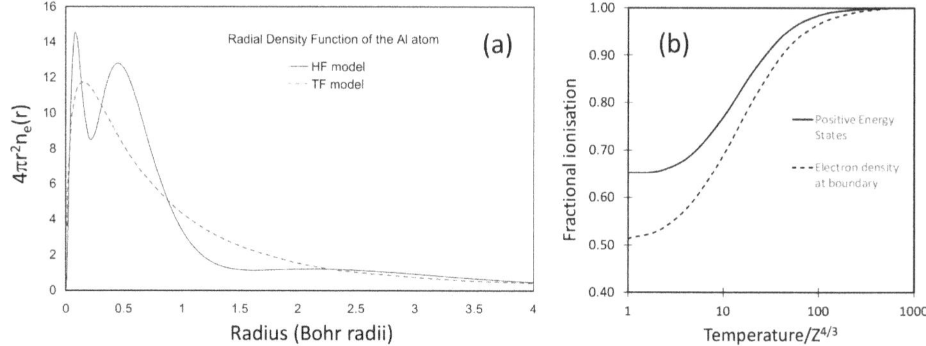

Figure 1.6. (a) Electron radial distribution in a Thomas–Fermi calculation for Al compared to a Hartree–Fock calculation (Code courtesy of G Gribakin, simulation by J Angulo-Gareta). (b) Comparison of fractional ionisation predicted by the TF model, by integrating over positive energy states (solid line) and taking the electron density at the ion-sphere boundary (dashed line). The scaled density is $\rho_0 = 0.167$ g cm^{-3} and for any given element we get this same curve at a density of $\rho = AZ\rho_0$.

The other method is to consider the number of states with positive energy; $p^2 > 2m_e eV$. This is obtained by taking the integral in equation (1.33) from a lower limit of $(2m_e eV)^{1/2}$ instead of zero or in equation (1.34) the incomplete Fermi–Dirac integral, $F_{1/2}(x, \alpha)$ can be used where instead of the integral being from zero to infinity, the lower limit is given by;

$$\alpha = \left| \frac{eV(r)}{k_B T} \right| \tag{1.39}$$

The average ionisations used in figure 1.5 were obtained using this latter method. These two methods agree well at high temperatures but can diverge fairly significantly at lower temperatures as discussed by More and others [64, 65]. In figure 1.6(b) we see a calculation of how ionisation fraction varies with temperature for both methods. The convenient scaling of the TF model has been used to generate an analytical approach to giving average ionisation that can be readily implemented in a hydrodynamics code and solved in-line [64]. In this, the input parameters are mass density, temperature, atomic number and atomic mass number.

Mixtures in the Thomas–Fermi model

The discussion above has assumed the Thomas–Fermi model is used for a single material. In practice we will often carry out experiments on a sample that is a mixture of elements. The Thomas–Fermi approach can be adapted to account for this, for example [66], by adopting a solution that iteratively adjusts the ion-sphere boundaries of multiple components until the chemical potential calculated at each ion-sphere boundary is equalised. The pressure for each element should then be the same and the fractional ionisation can be calculated for each element using the electron density at the boundary, in the same way as above.

Another, perhaps more rapid, approach has been to utilise the scaling properties of the TF model. For a given material, we know the composition and, for our target density, we can make an estimate of the partial density for each element and calculate the pressure for the common temperature, e.g. [67]. We can then find the average pressure of the different components and scale the density for each element to match this average pressure. This will give us a new total density that may not match the target density and so each element's partial density is then scaled to give the correct density and the process is repeated iteratively to a final solution where each element gives the same common pressure. In order to make this iteration rapid, we can use tabular results for each element that give the pressure and average ionisation over a grid of temperature and density values that can be interpolated and used for whatever mixture is chosen.

1.4.2 Modelling an EOS

An approach that is different from building a look-up table is to build an analytical model that can be relatively easily implemented and will run quickly. This approach has been taken, for example, by More *et al* [67] who start with an assumption that the Helmholtz free energy can be treated in an additive way;

$$F(\rho, T_e, T_i) = F_i(\rho, T_i) + F_e(\rho, T_e) + F_b(\rho, T_e) \tag{1.40}$$

where the contribution for ions, F_i and electrons, F_e, are treated separately and F_b is a term that can account for electron exchange effects including bonding in solids. Once this has been achieved, thermodynamic relationships can be used to derive pressure, entropy, and internal energy from the input values of density and temperature, for example;

$$P = \rho^2 \frac{\partial F}{\partial \rho} \tag{1.41}$$

$$S = -\frac{\partial F}{\partial T} \tag{1.42}$$

$$E = F + TS \tag{1.43}$$

where the energies, F and E are per unit mass. The Helmholtz free energy is given by $F = -k_B T \ln Z(T, \rho)$, where $Z(T, \rho)$ is the partition function, which is, in general, density and temperature dependent. It is beyond the remit of this book to derive the Helmholtz free energy but expressions are given in some important limits, in for example [67]. In the approach of More *et al*, the contribution from ions is treated in a variety for ways, depending on the expected phase. At low density and high temperature, an ideal gas model is used. By contrast, high density phases are handled by a variety of laws. The electrons are treated in the Thomas–Fermi approximation as above, whilst, for example, the Lindemann melting law is used to decide if we have a solid or fluid phase, the Grüneisen pressure law is used to decide ion thermal pressure and, if there is a solid phase, then the treatment will depend on the temperature relative to the Debye temperature.

1.4.3 Mie–Grüneisen equation of state

Although, it is really more appropriate to shocked solids than WDM, it is worth a short mention of the Mie–Grüneisen equation of state, as it is widely used in high pressure shock physics and does come into use in WDM relevant experiments. Furthermore, as part of EOS such as QEOS, it can form part of radiation-hydrodynamics simulations of matter that is brought from ambient to WDM conditions. There are different approaches to calculating the Grüneisen parameter, but we can start by defining it thermodynamically by;

$$\gamma_G = V\left(\frac{\partial P}{\partial E}\right)_V \tag{1.44}$$

where V is specific volume, P is pressure and E is internal energy. For a perfect gas with $P = nk_BT$ and the internal energy $E = 3nk_BT/2$, we can see that $\gamma_G = 2/3$. For a Debye solid, thermodynamic relationships can be used to show that;

$$\gamma_G = \frac{\partial \log(\Theta_D)}{\partial \log(\rho)} \tag{1.45}$$

where Θ_D is the Debye temperature and ρ is the density. Values reported for metals range from about 0.9 for Li to 2.6 for lead [68]. It is normally assumed that the Grüneisen parameter is density dependent only;

$$\gamma_G(\rho) = \gamma_G(\rho_0)\left(\frac{\rho_0}{\rho}\right)^a \tag{1.46}$$

where, for example Kraus *et al* [69] give $\gamma_G(\rho_0) = 1.35$ and $a = 2.6$ for stishovite, an important mineral in geophysics. The Mie–Grüneisen pressure law then gives;

$$P_i = \gamma_G(\rho)\rho E_i \tag{1.47}$$

where we have the pressure P_i and internal energy E_i for the ions. Obviously, we are not dealing with a Debye solid as our WDM sample, but equation (1.44) is quite general and calculating a Grüneisen parameter for liquid and melting phases has been carried out [70, 71]. We shall see the Grüneisen parameter again in chapter 5, when we deal with temperature measurement in shock release.

1.5 Creating and probing warm dense matter

Amongst the general requirements for the creation of WDM, we can consider the need for a uniform sample, as discussed for example, by Ng *et al* and Forsman *et al* [6, 72] who discuss the notion of an idealised WDM slab. This is important because strong gradients in the density can mask the physics we are trying to probe and make comparison with simulation and theory problematic. For this reason, direct heating of a solid by laser-irradiation is more usually ruled out as a useful method, except for ultra-thin foils. Otherwise, whilst high intensities lead to a hot plasma on the surface and strong electron thermal conduction inwards from the laser-plasma can create

WDM conditions in the solid, this is inevitably accompanied by very strong density and temperature gradients. The use of very high intensity, short pulse irradiation to create a population of super-thermal electrons that can penetrate a solid and create WDM is a possible exception that will be discussed in a later chapter. Of course, we may sometimes be interested in non-steady state situations where heat flows and currents may be generated by gradients, but resolving those gradients in itself becomes a challenge to be met for the diagnostics employed.

As well as spatial uniformity and resolution, we also need to consider the timescale on which we can create WDM samples. As we noted earlier, static methods such as DAC only probe a limited part of the WDM regime and we need to employ dynamic methods, such as volumetric x-ray heating or shock compression, where our sample is rapidly heated and/or compressed and maintains WDM conditions due to mass inertia for a period suitable for probing. Consider, for example, if we heat a sample to a few eV at solid density. The speed of sound, c_s, can be expressed as;

$$c_s = \sqrt{\frac{\gamma_a P}{\rho}} \tag{1.48}$$

where P is the pressure, γ_a is the adiabatic index and ρ is the mass density and we have assumed that the electrons provide the thermal pressure and are isothermal. Taking an example of a thin foil of thickness, $d = 20$ nm, heated to around 10 eV by direct heating with a laser and containing $\sim 2 \times 10^8$ J kg^{-1} deposited energy at solid density [6], we can estimate a sound speed of about 2×10^4 ms^{-1} and thus a disassembly time, $\tau = d/c_s$ of 1 ps. This is very short and diagnostics of such samples are often based on short pulse optical probing of the surface in order to explore the conductivity of the sample, e.g. [73]. Depending on the material, we will generally estimate, in WDM conditions, a sound speed of similar magnitude.

If we look at experiments carried out with larger samples, perhaps of mm-scale, this implies that decompression will occur on a timescale of 10^{-8} to 10^{-7} s. There are many diagnostics that can work well on this timescale, as we shall see, and this points to a clear advantage in using larger samples. However, as we shall also see, larger samples themselves can present problems. The higher opacity to x-rays can inhibit the creation and probing of a large sample. This is especially true for mid to higher Z elements and, in many cases, we need to use smaller samples and deal with the faster target evolution. In addition to this, we can note that a typical WDM sample pressure of 100 GPa, means an ~ 1 mm^3 volume requires ~ 100 J of energy to be deposited within the sample. This implies the need for large energy facilities, some of which are discussed in chapter 6. Needless to say, the timescale for heating of the sample is also a key parameter here and needs to be short enough that time resolved diagnostics can probe a sample that does not evolve excessively during the probing time. The tolerance to target evolution will vary from experiment to experiment but the main criterion will be the degree to which temporally averaged values of the parameter measured will allow discrimination between competing theoretical models.

1.5.1 Equilibration timescales

The timescale for heating is also important from the point of view of the microscopic equilibration of the sample. We have noted that the understanding of planetary interiors is a prime motivation for WDM research. Such interiors may have gradients and dynamic processes but these are generally expected to be on a much longer timescale than those in our experiments and we wish to study matter under more or less thermally and hydrodynamically equilibrated conditions. If experiments are to be relevant, we need to concern ourselves with the rates at which melting, electron–ion energy exchange and phase changes occur.

For experiments where we directly heat a thin slab with an optical laser, we have already indicated that, in order to measure the properties of a uniform slab, probing must take place on a sub-picosecond timescale. Since the phonon oscillation time in a solid is sub-picosecond [74] the energy deposition, probing and equilibration times are of similar magnitude and the rate of equilibration must be taken into account.

With the advent of free electron x-ray lasers that can deposit energy on a sub-picosecond timescale [1], this is an issue that will only become more important. In these cases, we note that the pulse duration is so short that it brings within our grasp the possibility of exploring the fundamental processes of melting and electron–ion equilibration that are involved in turning a solid into WDM. For example, Mazevet *et al* [75] have explored, via molecular dynamics simulations, the timescale on which the face-centred cubic lattice of solid density gold becomes disordered due to heating with an x-ray laser pulse. Depending on the energy deposition, this can take 10s of picoseconds and it is clear that x-ray diffraction experiments with a separate probe would be capable of following this decay by analysing the scattered radiation. It would be a mistake, however, to assume that equilibration effects are only likely to be only likely to be evident in experiments taking place on a picosecond timescale. As has been discussed by Ng *et al* [76], strong coupling may affect the timescale on which electron–ion energy exchange occurs in shocks driven by nanosecond lasers.

We have seen above how electron degeneracy plays a role in determining the thermal and electrical conductivity in a dense plasma due to Pauli blocking of possible final states for electrons. Similarly the timescale on which electrons and ions can exchange energy via collisions is also affected. Brysk [77] studied this and showed that the electron–ion equilibration time should be increased by a factor;

$$\frac{\tau_{\text{Brysk}}}{\tau_{\text{Spitzer}}} \approx 1 + \frac{4N_e\pi^{3/2}\hbar^3}{(2m_e k_B T_e)^{3/2}} \tag{1.49}$$

We can see easily that at low density and high temperature, the Spitzer–Brysk result recovers the classical limit. As an example, even in relatively dense plasma with electron density 10^{23} cm^{-3} and an electron temperature of 10 eV, we would have a factor of $\sim$1.13, whereas for shock compressed matter where the temperature may be of order 1 eV and the electron density close to 10^{24} cm^{-3} this rises to a factor of over 100.

We have seen how strong coupling affects the microscopic arrangement of the ions and thus various plasma properties such as the resistivity. This is connected to the fact that there is an effect on the electron–ion equilibration time. This is an important point because, in WDM experiments, we usually directly heat either the electrons, perhaps with x-ray heating or the ions, for example in shocks, and then we rely on rapid energy exchange to reach an equilibrium condition.

We can understand this simplistically in terms of the effective mass of the ions with which the electrons collide. We are used to the idea that in electron–ion equilibration the ratio of masses is important. This comes about since, if a particle of m_a collides with a larger particle, M_A, initially at rest, then for a wide range of collision angles, we can say that $|p_a| \sim |p_A|$ after collision. Since the kinetic energy in the non-relativistic limit is $p^2/2m$ we can see that the ratio of kinetic energies is given by m_a/M_A. If we have strong coupling between ions, then momentum taken up via collisions with electrons, is shared with neighbouring ions and thus the effective mass is increased and the rate of energy exchange reduced.

The role of strong coupling in determining the rate of energy exchange between electrons and ions in dense plasmas and WDM has been considered by several authors [78]. Often this is expressed as a generalised coupling constant such that the internal energy of the electrons is governed by;

$$\frac{\partial}{\partial t}(\rho E_e) = -\frac{\partial}{\partial x}\left[\rho u\left(E_e + \frac{P_e}{\rho}\right)\right] + -\frac{\partial}{\partial x}\left(\kappa\frac{\partial T_e}{\partial x}\right) + u\frac{\partial P_e}{\partial x} + \frac{\partial E_{in}}{\partial x} - g(T_e - T_i)\frac{\rho}{\rho_0} \quad (1.50)$$

where E_e is the internal energy density for the electron sub-system, E_{in} is an input source of energy to the electrons, κ is the thermal conductivity and ρ_0 is the initial mass density. In the last term, g is the coupling constant between electrons and ions.

In Dharma-Wardana and Perrot [79] it is shown that theoretical approaches accounting for coupled modes of ions lead to electron–ion coupling constants up to four orders of magnitude lower than the Spitzer–Brysk rate. Take, for example, the case of solid density Al, where the electrons are suddenly heated to 10 eV, as can be achieved by an x-ray free electron laser. If the ions are assumed to start at the melting temperature of Al, the Spitzer–Brysk model predicts a coupling constant of 0.3634×10^{20} W/K/m^3, whilst the coupled mode calculation of Dharma-Wardana and Perrot gives 0.2374×10^{16} W/K/m^3. This changes the expected equilibration time from being sub-picosecond to hundreds of picoseconds. Indeed, in the work of Ng *et al*, [76], study of the optical emission from the rear of a sample from which a shock wave has broken out indicates that the emission history can be understood if the equilibration time is indeed of order 100s picoseconds.

In chapter 2, we will consider shock generation and the melting point of a shock compressed sample will be discussed along with the possibility that the speed of melting is an issue and superheating may occur with samples retaining a solid state structure, whilst still being at above the equilibrium melting temperature for the applied pressure [80, 81].

1.5.2 Summary

We have seen that the challenge of theoretically describing WDM is affected by the issues of strong coupling, electron degeneracy and partial ionisation. We have also seen that the challenges in conducting and interpreting experiments include time-scale effects, both from the point of view of hydrodynamics and of microscopic equilibration processes.

In the following chapters we will discuss the generation and diagnosis of WDM by a variety of means, addressing generation and diagnosis together, in some cases, as some experiments lend themselves to particular diagnostic techniques. We will consider both lasers and ion beams as tools for WDM creation but we will also seek to broadly categorise techniques as shock/ramp compression and volumetric heating, both of which can be achieved by more than one technique.

References

[1] Lee R W *et al* 2003 *J. Opt. Soc. Am.* B **30** 770
[2] Koenig M *et al* 2005 *Plasma Phys. Contr. Fusion* **47** B441
[3] Graziani F, Desjarlais M P, Redmer R and Trickey S B (ed) 2014 *Frontiers and Challenges in Warm Dense Matter* (Berlin: Springer)
[4] More R, Yoneda H and Morikami H 2006 *J. Quant. Spectrosc. Radiat. Transfer* **99** 409–24
[5] Nettelmann N, Redmer R and Blaschke D 2008 *Phys. Part. Nucl.* **39** 1122
[6] Ng A, Ao T, Perrot F, Dharma-Wardana M W C and Foord M E 2005 *Laser Part. Beams* **23** 527
[7] Renaudin P, Blancard C, Clérouin J, Faussurier G, Noiret P and Recoules V 2003 *Phys. Rev. Lett.* **91** 075002
[8] Lee R W, Kalantar D and Molitoris J 2004 UCRL-TR-203844
[9] Mattern B A and Seidler G T 2013 *Phys. Plasmas* **20** 022706
[10] Dharma-Wardana M W C 2006 *Phys. Rev.* E **73** 036401
[11] Perrot F and Dharma-Wardana M W C 1993 *Phys. Rev. Lett.* **71** 797–800
[12] Davidson R *et al* 2003 *Frontiers in High Energy Density Physics: The X-Games of Contemporary Science* (Washington, DC: The National Academies)
[13] Falk K 2018 *High Power Laser Sci. Eng.* **6** e59
[14] Johnson J A 2009 *Publ. Astron. Soc. Pac.* **121** 309–15
[15] Bowler Brendan P 2016 *Publ. Astron. Soc. Pac.* **128** 102001
[16] Wright J T *et al* 2011 *Publ. Astron. Soc. Pac.* **123** 421–2
[17] Perryman M 2006 *The Exoplanet Handbook* (Cambridge: Cambridge University Press)
[18] Ross M 1981 *Nature* **292** 435
[19] Benoit M, Bernasconi M, Focher P and Parrinello M 1996 *Phys. Rev. Lett.* **76** 2934–6
[20] Cavazzoni C, Chiarotti G L, Scandolo S, Tosatti E, Bernasconi M and Parrinello M 1999 *Science* **283** 44–6
[21] Millot M, Coppari F, Ryan Rigg J, Correa Barrios A, Hamel S, Swift D C and Eggert J H 2019 *Nature* **569** 251–5
[22] Nettelmann N, Holst B, Kietzmann A, French M and Redmer R 2008 *Astrophys. J.* **683** 1217–28
[23] Nettelmann N, Becker A, Holst B and Redmer R 2012 *Astrophys. J.* **750** 52

[24] Buccino D R, Helled R, Parisi M, Hubbard W B and Folkner W M 2020 *J. Geophys. Res.: Planets* **125** e2019JE006354
[25] Dubrovinsky L S, Saxena S K, Tutti F, Rekhi S and LeBehan T 2000 *Phys. Rev. Lett.* **84** 1720
[26] Yoo C S, Akella J, Campbell A J, Mao H K and Hemley R J 1995 *Science* **270** 1473
[27] Wang J *et al* 2013 *J. Appl. Phys.* **114** 023513
[28] Riley D 2018 *Plasma Phys. Control. Fusion* **60** 014033
[29] Kohanoff J 2006 *Electronic Structure Calculations for Solids and Molecules* (Cambridge: Cambridge University Press)
[30] Hansen J P 1979 *Proceedings of the 20th Scottish Universities Summer School in Physics* (Edinburgh: SUSSP Publications)
[31] Brush S G, Sahlin L and Teller E 1966 *J. Chem. Phys.* **45** 2102
[32] Hansen J P 1973 *Phys. Rev.* A **8** 3096–109
[33] Wünsch K, Vorberger J and Gericke D O 2009 *Phys. Rev.* E **79** 010201(R)
[34] Vorberger J and Gericke D O 2013 *High Energy Density Phys.* **9** 178–86
[35] Hansen J-P and McDonald I R 2006 *Theory of Simple Liquids* 3rd edn (New York: Elsevier)
[36] Ornstein L and Zernike F 1914 *Proc. Acad. Sci.* **17** 793
[37] Ziman J M 1961 *Philos. Mag.* **6** 1013
[38] Ichimaru S 1982 *Rev. Mod. Phys.* **54** 1017–59
[39] McDougall J and Stoner E C 1938 *Phil. Trans. R. Soc.* **237** 67
[40] Latter R 1955 *Phys. Rev.* **99** 1854–70
[41] Cowan B 2019 *J. Low Temp. Phys.* **197** 412–44
[42] Rose S J 1988 *Proceedings of 35th Scottish Universities Summer School in Physics* (Edinburgh: SUSSP Publications)
[43] Clayton D D 1968 *Principles of Stellar Evolution and Nucleosynthesis* (Chicago: University of Chicago Press)
[44] Cox J P and Guili R T 1968 *Principles of Stellar Structure* (New York: Gordon and Breach)
[45] Spitzer L and Harm R 1953 *Phys. Rev.* **89** 977
[46] Kremp D, Schlanges M and Kraeft W-D 2005 *Quantum Statistics of Nonideal Plasmas* (Berlin: Springer)
[47] Gregori G, Glenzer S H and Landen O L 2003 *J. Phys. A: Math. Gen.* **36** 5971–80
[48] Fletcher L B *et al* 2015 *Nat. Photon.* **9** 274
[49] Greim Hans R 1997 *Principles of Plasma Spectroscopy* (Cambridge: Cambridge University Press)
[50] Wigner E and Huntington H B 1935 *J. Chem. Phys.* **3** 764
[51] Saumon D and Chabrier G 1989 *Phys. Rev. Lett.* **62** 2397–400
[52] Norman G E and Saitov I M 2019 *Contrib. Plasma Phys.* **59** e201800182
[53] Stewart J C and Pyatt K D 1966 *Astrophys. J.* **144** 1203
[54] Ciricosta O *et al* 2012 *Phys. Rev. Lett.* **109** 065002
[55] Hoarty D J *et al* 2013 *Phys. Rev. Lett.* **110** 265003
[56] Crowley B J B 2014 *High Energy Density Phys.* **13** 84–102
[57] Rosmej F B 2018 *J. Phys. B: At. Mol. Opt. Phys.* **51** 09LT01
[58] Ecker G and Kröll W 1962 *Phys. Fluids* **6** 62–9
[59] Iglesias C A and Sterne P A 2013 *High Energy Density Phys.* **9** 103–7
[60] Lyon S P and Johnson J D 1992 Sesame: The Los Alamos national laboratory equation of state database *LANL Technical Report* LA-UR-92-3407

[61] More R M 1994 *Laser Part. Beams* **12** 245–55
[62] Feynman R P, Metropolis N and Teller E 1949 *Phys. Rev.* **75** 1561–73
[63] Salzmann D 1998 *Atomic Physics in Hot Plasmas* (Oxford: Oxford University Press)
[64] More R M 1981 Atomic physics in inertial confinement fusion *Report* UCRL 8499
[65] Ying R and Kalman G 1989 *Phys. Rev.* A **40** 3927–50
[66] Shemyakin O P, Levashov P R and Krasnova P A 2019 *Comput. Phys. Commun.* **235** 378–87
[67] More R M, Warren K H, Young D A and Zimmerman G B 1988 *Phys. Fluids* **31** 3059–78
[68] Hasegawa M and Young W H 1980 *J. Phys. F: Met. Phys.* **10** 225–34
[69] Kraus R G *et al* 2012 *J. Geophys. Res.* **117** E09009
[70] Jeanloz R 1979 *J. Geophys. Res.: Solid Earth* **84** 6059–69
[71] Huang H-J, Jing F-Q, Cai L-C and Bi Y 2005 *Chin. Phys. Lett.* **22** 836–8
[72] Forsman A, Ng A, Chiu G and More R M 1998 *Phys. Rev.* E **58** R1248
[73] Price D F, More R M, Walling R S, Guethlein G, Shepherd R L, Stewart R E and White W E 1985 *Phys. Rev. Lett.* **75** 252–5
[74] Ashcroft N W and Mermin N D 1976 *Solid State Physics* (Philadelphia, PA: Saunders College Publishing)
[75] Mazevet S, Clérouin J, Recoules V, Anglade P M and Zerah G 2005 *Phys. Rev. Lett.* **95** 085002
[76] Ng A, Celliers P, Xu G and Forsman A 1998 *Phys. Rev.* E **52** 4299
[77] Brysk H 1974 *Plasma Phys.* **16** 927–32
[78] Chiu G, Ng A and Forsman A 1997 *Phys. Rev.* E **56** R4947–50
[79] Dharma-Wardana M W C and Perrot F 1998 *Phys. Rev.* E **58** 3705–18
[80] Luo S-N and Ahrens T J 2004 *Phys. Earth Planet. Inter.* **143–144** 369–86
[81] White S *et al* 2020 *Phys. Rev. Res.* **2** 033366

Chapter 2

Shock and ramp compression

2.1 General background

The generation of shock waves by the rapid application of high pressure is a subject that has been researched in detail for over 70 years [1, 2] and has been thoroughly described very well by other authors [3–7]. Much of this early research has concerned shocked matter that is still in the solid state, although it may have changed phase, for example iron changes phase from body-centred cubic to hexagonal close packed at a relatively modest shock pressure of around 13 GPa [2]. Since we are interested in warm dense matter (WDM), we are still primarily concerned with driving shocks into solid targets, but at high enough pressure to heat the sample beyond the melt temperature, which, itself, is pressure dependent. Thus, in this book, we will generally limit ourselves to a review of some important points judged relevant to the WDM experimentalist, although, of course, the temperature needed to melt the sample at the pressure of the shock is an issue of relevance to us. For our purposes, we will observe that a shock wave generates a sudden jump in the pressure, density, and temperature of the sample.

We can derive some conservation laws governing the propagation of a shock as follows. Imagine we have a piston moving into matter, initially at rest, driving a shock wave ahead of it, as in figure 2.1. The densities before and after compression are ρ_0 and ρ. The piston moves at velocity u_p. A shock is driven at speed u_s into the undisturbed matter ahead of the piston. The mass per unit area of material, compressed by the shock in a short time, is given by $\rho_0 u_s \Delta t$. This should be equal to the mass per unit area of the material between the moving piston and the shock front which is given by $\rho(u_s - u_p)\Delta t$. From this we get our first Rankine–Hugoniot equation;

$$\rho_0 u_s = \rho(u_s - u_p) \tag{2.1}$$

which is effectively a statement of conservation of mass. From this, we can see that if we can somehow measure both the particle speed and the shock speed we can infer

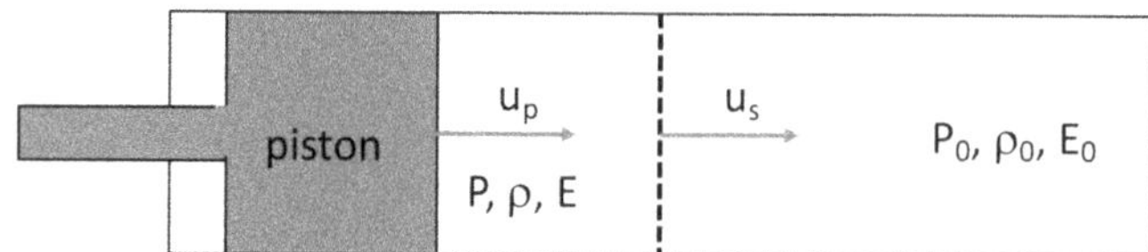

Figure 2.1. Schematic of a piston driving into a sample at speed u_p and driving a shock at speed u_s.

the compression. Next, let us consider the conservation of momentum imparted to the shocked material. Since the pressure is the force per unit area, the change in momentum (impulse) due to a force, acting for a short time at the shock front, is given by $(P - P_0)\Delta t$. This is then equal to the mass per unit area of material shocked, $\rho_0 u_s \Delta t$, multiplied by the speed at which it is set in motion, u_p. Thus, we arrive at our second equation;

$$P - P_0 = \rho_0 u_s u_p \tag{2.2}$$

So, we can see that measurement of the shock and particle velocity, not only gives the shock density but the pressure as well. In WDM work, we are working in a regime where the shock pressures are of order >100 GPa and so we would usually be fine in approximating;

$$P = \rho_0 u_s u_p \tag{2.3}$$

Finally, we can look at energy conservation. From our discussion of equation (2.1), and recalling that we are speaking of a unit area of shock, we can see that, during a small increment in time, there is a change in volume of the material compressed of $u_p\Delta t$. Thus, the PdV work done by a piston in compressing the material can be calculated as $Pu_p\Delta t$. The work done supplies the kinetic energy for setting the mass of material in motion as well as the change in internal energy in the compressed matter;

$$Pu_p = \rho_0 u_s \left[(E - E_0) + \frac{u_p^2}{2} \right] \tag{2.4}$$

where the overall units on each side are in terms of energy per unit area per unit time and E is in energy/unit mass. Assuming the initial conditions for the sample are known, there are five unknowns. These are the pressure, P, density, ρ, internal energy, E, of the shocked matter, the velocity of the shock, u_s and the particle velocity of the matter after it has been shocked, u_p. If we measure any two of these parameters we can, in principle, solve for the others. These three equations above can be combined and rearranged to give the Rankine–Hugoniot relation;

$$E - E_0 = \frac{(P + P_0)(V_0 - V)}{2} \tag{2.5}$$

where $V_0 = 1/\rho_0$ and $V = 1/\rho$. For each value of shock pressure, there is a unique increase in internal energy and density of compression and the locus of points can be plotted out on the shock Hugoniot. The above discussion has centred on the conditions either side of a shock front. It is worth making a brief mention of why a shock forms In brief, the sound speed, c_s, for a solid, is related to the isentropic bulk modulus, K_s via;

$$c_s = \sqrt{\frac{K_s}{\rho}} \tag{2.6}$$

Since, for almost all materials, K_s increases with pressure, this leads to a sound speed that increases with pressure, leading to the situation in figure 2.2(b) where we see an essentially arbitrary pressure profile steepens into a shock as the sound speed of the disturbances at higher pressure catch up with the front edge of the pressure wave. A similar situation arises in shocked gases [3], where the pressure takes the place of the bulk modulus in equation (2.6).

For a shock in a given material at a given pressure, there is not only a characteristic specific volume but a characteristic shock speed, particle speed and temperature so that, often, the Hugoniot is plotted using these different variables. It is important to note, at this point, that the Rankine–Hugoniot equations above do not give this temperature even if we solve by measuring two quantities such as the shock and particle velocity. What we get is the internal energy, E, and this has to be linked to the temperature by having a knowledge of the heat capacity as part of the equation of state for the material under shock compressed conditions. In chapter 5, we will look at some techniques deployed to measure shock temperatures. In many cases, however, we infer the temperature by using experimental measurements of other parameters such as shock velocity, in tandem with an equation of state that gives the temperature for a given shock pressure. An example of such an equation of state is the Quotidian equation of State (QEOS), of More *et al* [8], discussed in chapter 1, which has been implemented in codes used for simulation of shock compression. Figure 1.1 shows the Hugoniot curves, with temperature, for iron and hydrogen taken from the SESAME equation of state database [9]. It is, of course, important to remind ourselves that the

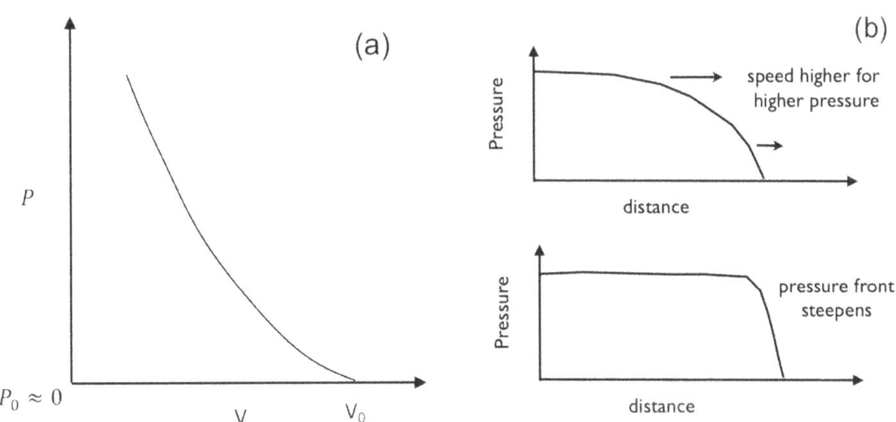

Figure 2.2. (a) Characteristic Hugoniot plot for an unspecified material. (b) Simple schematic of how sound speed increasing with compression leads a pressure profile to steepen into a shock front.

compressed matter does not evolve along these curves but, for a given shock pressure, jumps from the ambient condition to a point on the curve. As discussed, for example, in [10], for a weaker to moderate shock, the shock speed may be slower than an elastic wave and we may see two shocks, one of them an elastic precursor, for a detailed discussion see [11]. For WDM experiments, we are generally interested in shocks, where the pressure is great enough to complete melting of the sample and the shock speed is greater than the elastic wave speed, and so we will not discuss these situations further in this text.

In figure 2.2(a), we can see the pressure curve rising with increasing steepness as the sample is compressed. For a very strong shock, we can consider the well-known result that the maximum compression for an ideal gas is given by;

$$\frac{\rho}{\rho_0} = \frac{(\gamma_a + 1)}{(\gamma_a - 1)} \tag{2.7}$$

The value of the adiabatic index, γ_a, depends on the number of degrees of freedom, F, and is given by $\gamma_a = (F + 2)/F$ and so, if we assume our sample is effectively a monatomic gas with just three translational degrees of freedom, we have $\gamma_a = 5/3$ and the maximum compression is a factor of 4. Solids are, of course, much harder to compress than gases and shocks of well above 100 GPa are needed to achieve significant compression, as we shall see in examples below.

An important experimental observation in solids is that there is an almost linear relationship between the shock velocity, u_s and the particle velocity, u_p behind the shock front;

$$u_s = u_0 + bu_p \tag{2.8}$$

where, for example, with Fe, the constants are $u_0 = 3935\ \mathrm{ms}^{-1}$ and $b = 1.578$ [12] and the fit is good to well over 400 GPa, even though it is expected to undergo shock melting between pressures of 220–280 GPa. As discussed by Seigel [10], we might expect the value of u_0 to be the hydrodynamic sound speed. In fact, for many materials, u_0 is within 20% of the sound speed, as is the case for iron itself (sound speed 4.99 km s^{-1}). A short table of values for a range of materials is shown in table 2.1.

Table 2.1. Experimental fits to $u_s = u_0 + bu_p$ for several elements, taken from a larger table compiled in Eliezer [13].

Element	Density (g cm^{-3})	u_0 (km s^{-1})	b
Lithium	0.534	4.77	1.066
Aluminium	2.703	5.24	1.4
Copper	8.93	3.94	1.489
Tin	7.287	2.59	1.49
Tantalum	16.69	3.41	1.2
Tungsten	19.3	4.03	1.237
Gold	19.3	3.08	1.56

Using equation (2.1) we can use this linear relationship between shock and particle speeds to get;

$$\frac{\rho}{\rho_0} = \frac{u_0 + bu_p}{u_0 + (b - 1)u_p} \tag{2.9}$$

If we consider a stronger shock, as u_p gets larger we can see that a limiting compression ratio of $b/(b - 1)$ emerges and in the case of iron this should be ~2.7, whilst for several of the other cases, particularly, Li, it is predicted to be greater than the four times compression of a perfect gas. In table 2.1 we can see that there is a fair range in the values of u_0 and b. However, it has been shown that at high pressures, above around 1 TPa, metals show a so called *universal Hugoniot* with;

$$u_s(\text{km s}^{-1}) \simeq 5.9 + 1.22u_p \tag{2.10}$$

This was shown for elemental metals ranging from Al ($Z = 13$) to Mo ($Z = 42$) [14] but also for metal oxides such as $Gd_3Ga_5O_{12}$ [15] and indicates the creation of a universal fluid metallic state.

Secondary Hugoniots and ramp compressions

The principal Hugoniot that we have considered thus far applies when we start with a sample under what we consider to be ambient conditions, typically $P_0 = 0$, and $T \sim 300$ K. However, sometimes, we can carry out an experiment with different initial conditions. For example, diamond anvil cells (DAC) have been used to pre-compress both helium and hydrogen [16] to pressures in the GPa range.

As we can see in figure 2.3, this technique means we generate a different Hugoniot for each starting pressure. This allows a degree of independent control over the

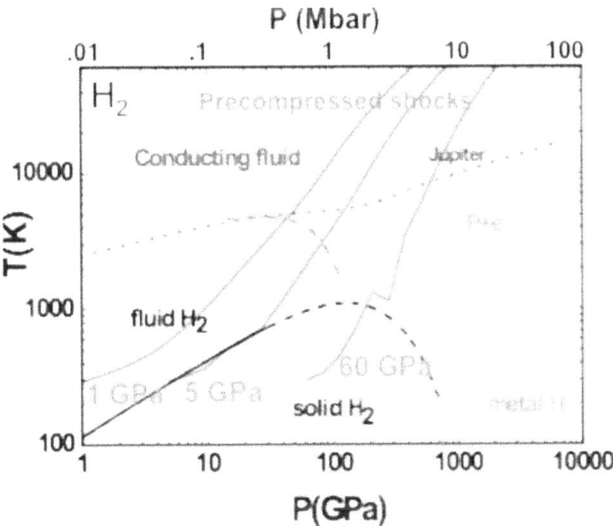

Figure 2.3. Calculated shock Hugoniots for hydrogen at a range of different starting pressures. Reproduced from [16], with the permission of AIP Publishing, ©AIP 2009.

pressure–temperature combination generated that would not be possible with only a single starting pressure; allowing us to better explore the isentrope of gas giants such as Jupiter. The principle behind this approach can be understood by considering the entropy, S, created by a single shock. For an ideal gas this is given by;

$$\Delta S = C_V \ln\left[\frac{PV^{\gamma_a}}{P_0 V_0^{\gamma_a}}\right] \tag{2.11}$$

This is always positive for a shock, which is an irreversible process. So, by choosing a different starting point, we generate different amounts of entropy for a given final shock pressure and, hence, different Hugoniots. In the discussion thus far, we have been considering different starting pressures for a single steep shock front, but we can also consider compression with multiple shocks. It can be shown that, for a weak shock, the discontinuity in entropy across the shock front is given by;

$$S - S_0 = \frac{1}{12T_0}\left[\frac{\partial^2 V}{\partial^2 P_0}\right]_S (P - P_0)^3 \tag{2.12}$$

We can see, then, that there is a cubic dependence on pressure. A consequence of this is that, if we wish to shock a sample to a given final pressure, $P > P_0$ then doing so in two stages, firstly to $P/2$ and then to P, generates four times less entropy than doing so in a single shock. Several authors have, in the past, explored such an approach with laser-driven shocks, both experimentally and in simulations [17–19], where coalescence of a strong shock with an earlier weaker shock, moving at slower speed, is used to generate compressions in excess of the single shock limit whilst restricting the temperature rise.

We can see the possibilities illustrated in figure 2.4, where we show a pair of hydrodynamic simulations carried out with the HYADES radiation-hydrodynamics code [20]. We take a 50 μm thick Al foil coated in 5 μm of CH as an ablator and drive shocks with 351 nm laser pulses that have a 0.2 ns rise and fall time, with a 1 ns flat top. In the first simulation, we have a single pulse with intensity 10^{14} W cm^{-2}.

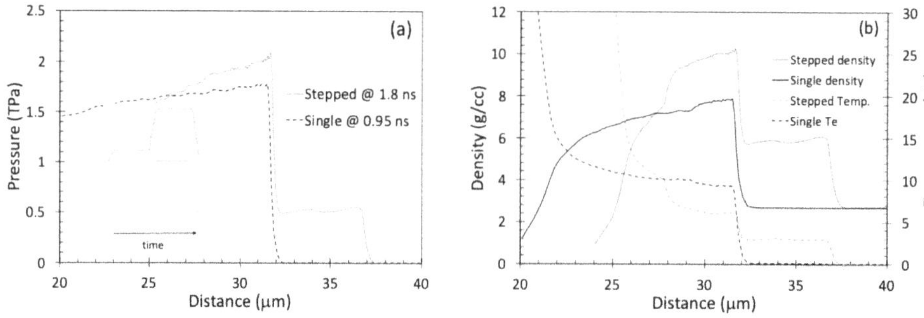

Figure 2.4. Simulated hydrodynamic profiles created with the HYADES code as described in the text. We show only the Al part of the target here. (a) As we can see, when the main shock front has reached near the middle of the foil, the pressure is roughly similar for both single and stepped shock cases. (b) Despite the similar pressure, the stepped target case has generated a higher density, lower temperature sample, away from the shock Hugoniot.

This generates a strong shock with shock speed ~33 km s^{-1} in the Al, that generates compression to nearly 8 g cm^{-3}, which is close to the compression estimated using the universal Hugoniot above. In the second simulation, we use a similarly shaped pre-pulse starting 1.2 ns earlier and with a lower intensity of only 2.5×10^{13} W cm^{-2}. This pre-pulse creates a weaker first shock (~18 km s^{-1}) with a compression that is, again, similar to the prediction of the universal Hugoniot. The second, stronger, shock catches up with the first and compresses the pre-shocked sample, reaching a density close to 10 g cm^{-3}, well in excess of the single shock Hugoniot limit. As we see, this is for a similar final shock pressure in the main drive pulse as for the single shock case, which is to be expected as we use the same peak laser intensity.

Today, pulse shaping technology [21] allows us to control, far more closely, the shape of the laser pulse driving the pressure applied to a sample. Early work on laser-driven fusion, e.g. [22, 23] showed that isentropic compression for a spherical capsule should use a 'concave' pressure history of form;

$$P(t) = P_0(1 - \tau^2)^{-2.5} \tag{2.13}$$

where $\tau = t/t_c$ and t_c is the time of collapse to a compressed core. Pulse shaping such as this can be used in WDM experiments to create a gentler ramp compression, closer to isentropic conditions and allows us to explore a wider range of conditions away from the shock Hugoniot [24, 25]. We should note, at this point, that in many cases isentropic compression will generate samples below the temperature that we would normally consider as being in the WDM regime. Nevertheless, Amadou *et al*, for example, [26] have used quasi-isentropic pulses to explore the melt-line for Fe at pressures up to 700 GPa, which is a regime relevant to the interior of potential 'super-Earth' exoplanets.

As we have already stated, for WDM studies we are generally interested in samples that are no longer in their crystalline form and an important aspect of shock heating is the determination of the point at which the sample will melt. As we have seen above, we expect a shock to heat the sample and the entropy generated increases rapidly with shock pressure. The temperature at which the sample melts will generally increase with pressure, following the so called *melt line*, but this increases at a slower rate than the rate at which increasing shock pressure raises the temperature. Thus, we expect that, at some shock pressure, the Hugoniot will cross the melt line, as illustrated in figure 2.5. There are several different criteria that can be used to decide at what point a solid will melt: for a review see [27]. The Lindemann criterion assumes that melting occurs when the average displacement of an atom from its lattice position reaches a particular fraction of the lattice spacing. This leads to a melt temperature as a function of density, ρ given by;

$$T_m(\rho) = \alpha\Theta_D^2/\rho^{2/3} \tag{2.14}$$

where Θ_D is the Debye temperature for the material and α is a material dependent constant. For pressures in the Mbar (100 GPa) regime, this dependence leads to significantly higher melting temperatures. For example, Fe starts to melt only when the pressure reaches ~220 GPa and the melting point is calculated to be around 6000 K, see for example [28], compared to 1811 K at ambient pressure. Melting can be looked

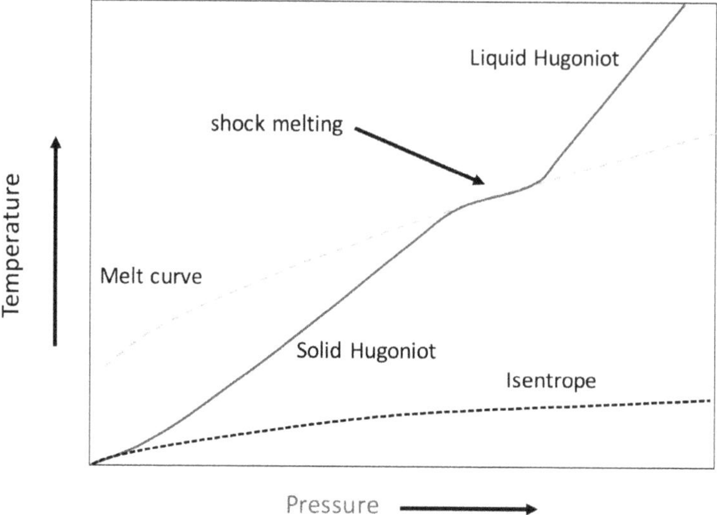

Figure 2.5. Schematic illustrating how isentropic compression compares to shock compression. By varying the rate at which compression occurs we can, in principle, explore the regime between the higher temperature shocked states and the isentropically compressed states.

for in shock experiments by its effect on the shock Hugoniot. The generalised Hugoniot we have shown in figure 2.2(a) is smoothly varying. If melting were present, this would show up as an inflection in the Hugoniot, as energy that would otherwise go into compression and heating is used to provide the latent heat for the melting process, as is illustrated schematically in figure 2.5. This could be observed, for example, by observations of the shock temperature as determined by optical emission measurements of the shock break-out in a series of experiments at increasing pressures spanning the regime where melting occurs [29]. Alternatively, measurements of the shock velocity as a function of pressure would also show an inflection where melting occurs, for example [30, 31]. It is worth pointing out that solid-solid phase changes will also show similar effects on the shock Hugoniot, though at a lower pressure and, for WDM research, we are usually more concerned with the melting of the sample.

In figure 2.5, we see how, as discussed above, the melt curve for a solid climbs with pressure but a shock above a given pressure will start to melt the sample, whereas an isentropic compression to the same pressure may leave the sample as a solid. For a short while, the Hugoniot follows the melt curve of the sample, as increasing pressure leads to a higher degree of melting until the melt is complete and we see the liquid Hugoniot at higher pressures. As noted earlier, in principle, the points at which melting starts and finishes can be determined by observation of the optical emission from the shock front, as a function of shock pressure. An interesting possibility is that a sample that has melted under shock compression can be re-solidified by further quasi-isentropic compression that does not raise the temperature to a significant further degree but nevertheless raises the melt temperature above the shock heated temperature, thus inducing solidification, though not necessarily to the original crystal structure. Such effects have in fact been observed in several materials, for example, see [32–34].

In this section we have already touched upon the use of intense, nanosecond duration, lasers focussed onto solid targets as way of generating shocks and ramp compressions, and indeed, this is one of the principal methods [35, 36]. In the next section we will consider the range of pressures accessible by this method and some of the experimental constraints we must consider. We will follow this with a discussion of other techniques for directly driving shocks, such as x-rays and ion beams. We will then discuss several methods for driving flyer plates to generate shocks and follow this with a discussion of impedance matching, which is an important technique in shock physics work for both flyer plates and directly driven shocks.

2.2 Directly driven shocks with intense lasers

In figure 2.6 we can see a simple schematic of a laser-driven shock with a solid target. The mechanisms by which a high intensity laser pulse creates a plasma when incident on a solid surface have been well discussed in the literature [37–39]. For laser-shock experiments, we are generally concerned with the regime where the laser wavelength, λ and intensity, I obey;

$$I\lambda^2 < 10^{15}\ \mathrm{W\ cm^{-2}} \tag{2.15}$$

In this regime, we expect the dominant absorption mechanism to be inverse bremsstrahlung (collisional absorption), which can be efficient [40]. A good reason for limiting ourselves to this regime is that, as the intense laser interacts with the plasma created at the solid surface, there are instabilities that can arise in the coronal plasma generated above the target. For example, filamentation or stimulated Raman scattering, which can both generate fast electrons that penetrate and pre-heat the solid target. This means that we no longer would have a good idea of the initial pre-shock target conditions making it more difficult to apply the Rankine–Hugoniot equations with any degree of certainty. As an example, the intensity threshold [41] for stimulated Raman scattering, in a coronal plasma of spatial scale-length, L, is;

$$I_{\mathrm{thresh}} > \frac{5 \times 10^{16}}{L^{4/3}\lambda^{2/3}}\ \mathrm{W\ cm^{-2}} \tag{2.16}$$

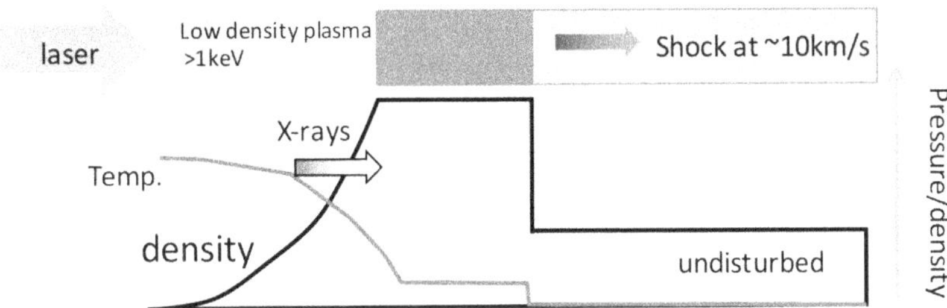

Figure 2.6. Schematic of a laser driven shock. The incident laser creates a hot plasma on the surface of a solid target. The hot plasma ablates from the surface, exerting a Mbar level pressure that drives a shock into the solid.

where L and the wavelength of the laser, λ are in microns. The coronal plasma can be typically at keV temperatures and expanding at a typical speed of 2×10^5 ms^{-1}, and so we expect $L > 200$ μm after 1 ns and thus the threshold is below 10^{14} W cm^{-2} for 527 nm laser light. The issue of pre-heating by supra-thermal electrons in shock experiments has been discussed by other authors, including Trainor and Lee [42]. The majority of the laser energy absorption, by inverse bremsstrahlung, occurs close to the critical density, which is given by [39];

$$n_c(\text{cm}^{-3}) \sim \frac{10^{21}}{\lambda^2(\mu\text{m})} \tag{2.17}$$

For a typical experiment, with optical lasers, the critical density is only a fraction of a percent of solid density. However, electron thermal transport carries energy into the target and creates a dense high temperature ablation surface where the pressure created drives a shock into the solid target.

2.2.1 Pressure generated

The subject of electron thermal transport in plasmas has generated an extensive literature but we will consider, here, a simple model that will help to estimate the expected shock pressure. This is based on the concept of the *free streaming limit*, where electrons in the plasma are assumed to flow in the same direction down the temperature gradient with the thermal velocity, given by $v_t = \sqrt{3k_BT_e/m_e}$. It has been common to model the electron heat flow as some fraction, f of the free streaming limit [43, 44], and we can equate the electron heat flow, Q_e to the rate of absorption of laser energy per unit area, I_{abs}, at the critical density to get;

$$I_{\text{abs}} = Q_e = fn_c v_t \frac{1}{2} m_e v_t^2 \tag{2.18}$$

By using equation (2.17) and the definition of thermal velocity, we can determine that, at critical density we expect;

$$T_e = 0.32\left[\frac{I_{\text{abs}}}{f}\right]^{2/3} \lambda^{4/3} \tag{2.19}$$

where T_e is in keV and I_{abs} is in units of 10^{14} W cm^{-2}. Taking typical upper values of $f = 0.1$, for example [45], $I_{\text{abs}} = 1$ and $\lambda = 0.53$ μm, we estimate $T_e \sim 0.6$ keV. If we then assume the pressure is applied at the critical density surface, we can use the perfect gas equation of state to derive;

$$P(\text{TPa}) = n_c k_B T = 0.24(I_{\text{abs}})^{2/3}\lambda^{-2/3} \tag{2.20}$$

and a pressure of around 0.4 TPa. This simple approach underestimates the shock pressure as the heat in fact flows upward above the critical density but it gives a good order of magnitude idea of what pressure we should expect. Experimentally, there are several scaling laws in the literature that describe the pressure of the shock as a function of the laser intensity and wavelength. For example, Thompson *et al* [46] give an empirical scaling;

$$P(\text{TPa}) = 0.8(I_{14})^{3/4}\lambda^{-1/2} \tag{2.21}$$

where the laser intensity is in units of 10^{14} W cm^{-2} and the wavelength is in microns. We can see that for the same conditions as our simple estimate, we can calculate pressures of order 1 TPa (10 Mbar). The scaling with intensity is in general agreement with the findings of other authors both in simulations and experiments, for example [47]. As can be seen, in equation (2.21), it is expected that a shorter wavelength will enhance the pressure. This is mainly a result of the fact that, for shorter wavelengths, the energy is deposited at a higher critical density where collisional absorption is more efficient and the laser energy is deposited closer to the ablation front. An additional advantage of shorter wavelength is that, as we can see from equation (2.16), the threshold for instabilities is higher for shorter wavelengths. This limits the chances that fast electrons will be generated that can penetrate into, and pre-heat, the sample.

The shock speed generated depends not just on the pressure but also on the target material, and is faster for lower density and generally in the regime of 10–30 km s^{-1}. This means that for a typical laser-drive duration of 1 ns, a sample of 10–30 μm thickness can be compressed before the pressure source is removed. This calculation leads us to a further consideration, that of radiative pre-heat.

At temperatures of order 1 keV, we can expect significant ionisation and, for mid-Z elements, we expect significant x-ray emission in the several-keV range. As with fast electrons, these x-rays can cause pre-heating of the sample before the shock compression. For example, the attenuation length of 5 keV photons is around 20 μm in aluminium [48]. This is of the order of the thickness we can drive a shock wave during the typical nanosecond pulse duration and so pre-heating can be a problem to overcome in many laser-driven experiments. One way in which this problem is mitigated is by the use of low-Z ablator on top of the sample material. The lower Z leads to less intense x-ray production. In some cases, a special pre-heat shield layer of higher Z material is used between the ablator and the sample (see [49]).

A lower Z, ablator, layer does not just reduce the x-ray pre-heating. In addition, the ablator also affects the shock loading history of the sample. The density mismatch leads to a mechanical impedance mismatch that causes multiple reflection of shock waves. If the ablator is a low density material such as parylene-N, $(C_8H_8)_n$, then the density is usually lower than that of the sample, meaning that the shock created in the ablator is reflected from the interface. The reflected shock in turn creates a release wave as it reaches the ablation surface. The release wave travels inwards towards the high-pressure region. This effect has been studied in detail, for example, by Swift and Kraus [50] and, for Al on Au, by Ng *et al* [51] where it was shown to generate an increase in the shock pressure but shortens the time over which the pressure is applied. We can see in figure 2.7 a comparison of the pressure history for two cases, one is Fe directly irradiated by a laser and the other is Fe coated with an ablator layer of CH. We can see that for the case with a CH layer, we reach a

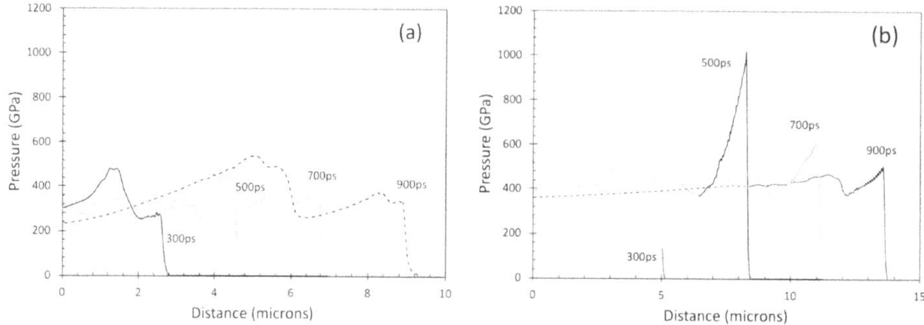

Figure 2.7. (a) Pressure profile for a bare Fe target irradiated by a 527 nm laser. The pulse rises in 300 ps assumes a flat top at 2×10^{13} W cm^{-2} for 1 ns and then falls over 300 ps. The Fe is 10 μm thick initially. (b) The same laser is used to irradiate a 5 μm layer of CH on top of a 10 μm thick Fe foil.

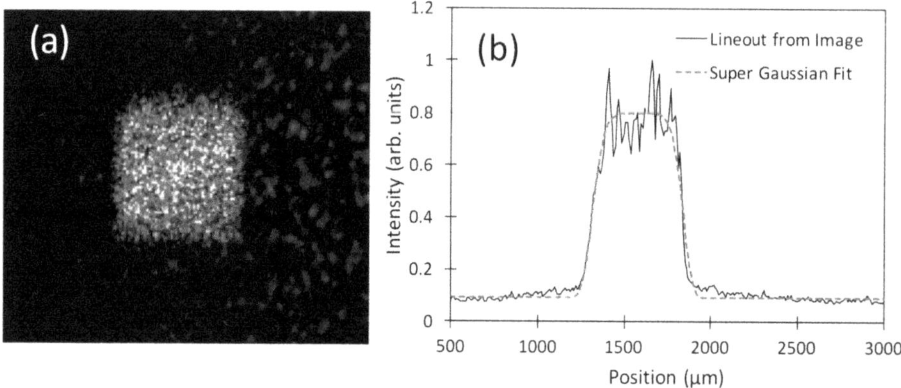

Figure 2.8. Example of a focal spot shape possible with phase plate technology. (a) Square focal spot with nominal 0.5 mm sides. A low level plateau around the focal spot is due to higher order diffraction and imperfection in manufacture. (b) Line-out average across central part of the spot with a super-Gaussian fit.

shock pressure nearly twice as high as the bare Fe case. However, this peak pressure decays rapidly, whereas in the bare Fe case, the pressure decays more slowly and has a flatter profile across the sample.

2.2.2 Focal spot uniformity

An important experimental consideration, in the generation of laser-driven shocks, is the uniformity of the illumination. Ideally, a flat-topped intensity distribution over a large focal spot (radius > 100 μm) is desirable in order to produce a shock that can be readily compared to one-dimensional hydrodynamic simulations for sample thicknesses of 10s microns. An important solution to this issue is the use of optical devices such as random phase plates [52] that can be readily placed in the laser beam-line. These have been available since the mid-1980s and there are now several varieties of such devices [53]. In figure 2.8, we not only show how a relatively flat-topped focal spot can be generated but also how the shape of the focal spot does not have to be circular.

The basic principle on which they work is that the phase plate is divided into many small regions where a deposited resist introduces optical path length changes that cause an effective phase change in that part of the beam of either π or 0 radians, which is assigned randomly. The effect is to break the beam into many beamlets that overlap on target creating a high frequency speckle pattern within a smooth overall envelope.

The speckle size depends on the diffraction limited spot size for the lens which, in turn, depends on the focal length, f_l of the lens, the beam diameter, Φ and the laser wavelength, λ. For a Gaussian beam profile, a lens will produce an Airy pattern at focus with a diameter (defined by the first minimum) given by;

$$d = 2.44 f_l \lambda / \Phi \tag{2.22}$$

Typically, it is of order microns. For example, for a 10 cm beam with 0.527 nm radiation and a 1 m focal length, $d = 12.8$ μm and this is the speckle size expected. For a nanosecond duration pulse, a coronal plasma of many 10s microns can be formed and the critical density surface is offset from the ablation surface where the maximum pressure is applied. Thus, for typical laser pulses of nanosecond duration, we expect that the high frequency speckle will be smoothed out via lateral thermal transport in the coronal plasma created during irradiation.

The size of the elements into which the phase plate is divided is inversely proportional to the size of the overall focal spot produced and can also be determined from equation (2.22). If we wish to have a focal spot of 100 μm, with a typical 527 nm wavelength, f_l = 1 m, beam, then application of equation (2.22) suggests that the element size needs to be ~12 mm. For a given focal length, a larger beam will have more elements and thus more statistical averaging, leading to a smoother irradiation profile. Another advantage for larger systems is that a larger focal spot can be used for a given irradiance and the ratio of speckle size to focal spot is smaller. The use of a larger focal spot also allows experiments to be more one-dimensional in nature. Thus, generally speaking, higher energy systems (>100 J) are to be preferred for this work.

There are other methods of beam smoothing that can be considered, such as induced spatial incoherence (ISI) [54, 55] This technique is different in that it does not rely on a fixed speckle pattern, but causes the speckles to fluctuate on a rapid timescale. ISI requires the use of a broad bandwidth laser where the coherence time of the pulse is given by;

$$\tau_c = \frac{1}{\Delta \nu} \tag{2.23}$$

If we place an echelon-like mirror in the beam, we can break the beam up into multiple beamlets which are each delayed relative to each other. As with random phase plates, the beamlets instantaneously create an interference pattern when they overlap in the focus of the lens. However, if the delays between beamlets are larger than the coherence time, which can be sub-picosecond, then the interference pattern fluctuates to smooth out the interference pattern. Another beam smoothing system, smoothing by spectral dispersion (SSD), works in conjunction with a phase plate.

The spectrally dispersed broad band laser pulse is incident on a phase plate, where the objective would be to have each element illuminated with a different frequency. The speckle pattern created by the interfering beamlets then shifts on a rapid timescale that depends on the bandwidth.

Spot size and one-dimensionality

The issue of how close to one-dimensional a shock wave is relates to any kind of shock experiment, but is particularly relevant to laser driven shock work, where, in order to access the TPa regime, we make use of focal spots less than 1 mm in diameter. In order to maintain a 1-D shock we must consider how thick the target can be. One criterion pointed out by Eliezer [13] assumes that, at the edge of the focal spot driving the shock, a rarefaction wave propagates inwards at the sound speed of the shocked region, c_s. If the shock travels at a speed u_s, then the time it reaches the rear of a foil of thickness, d, will be $t = d/u_s$ and the rarefaction will travel inward a distance given by $c_s d/u_s$. The condition for maintaining the planarity of the shock, at least in a large central region, is that the focal spot radius, $R_L > 2c_s d/u_s$, and since the sound speed of the shocked material is higher than the shock speed, this is approximated as $R_L > 2d$. This criterion is useful for all types of shock experiment.

One point that cannot be neglected, for laser-driven shocks, is that even if we have a nominal focal spot that is several times greater than the thickness of the shocked sample, the coronal plasma created by the laser-target interaction will have a finite size and may not expand in a one-dimensional way, thus affecting the coupling of the laser energy into generating pressure. For the case in figure 2.7, for example, the critical density surface moves away from the ablation surface at a speed of approximately $2 \times 10^5\ \mathrm{ms}^{-1}$, and by the temporal peak of the pulse, we see the critical density is 100 μm away. This means that we need a focal spot several times this size in order to maintain a 1-D behaviour for the problem as a whole. Alongside the dependence of laser-plasma instabilities on scale-length, the gradually increasing distance between the critical density and solid may place constraints on the length of laser pulse we can use to effectively drive a shock through a thick target.

In this section, we have discussed the generation of shocks with intense lasers. In section 2.6, we will discuss impedance matching experiments, which are an important way of accessing the equation of state for shock compressed materials. Before that, it will be useful to discuss some other shock generation methods, as they are also used in impedance match experiments.

2.3 X-ray driven shocks

An alternative to directly driving a shock with an intense laser is to use the laser to first create an intense flux of x-rays in the sub-keV to keV range and to use this to create a shock. This can have advantages in improved spatial smoothness of the shock drive and reduction of pre-heat, provided the sample is well designed. This approach is possible since, as we shall see further in chapter 3, intense lasers incident on high Z materials can be efficient sources of x-rays, especially in the sub-keV regime with high conversion efficiency of over 50% [56–58]. For sub-keV x-rays the absorption length in a solid is generally of micron or sub-micron size (for example, it is only 0.5 μm for 500 eV photons in Al) and so they can be used to ablate the sample surface, exerting an ablation pressure which, in turn, drives a shock wave. Pakula and Sigel [59, 60] have given the shock pressure generated for a self-similar, ablative, radiation driven, heat wave as;

$$P(\mathrm{TPa}) = 3.5C^{3/26}I^{10/13}t^{-3/26} \tag{2.24}$$

where we have a constant flux, I, in units of 10^{14} W cm^{-2} and t is in nanoseconds after the start of the flux. The constant, C, depends on the material and is given as ~7 for Au. An effective way to create the intense x-ray source is by irradiation of the inside of a mm or sub-mm sized gold cavity (hohlraum), with intense laser beams, as illustrated in figure 2.9. The emission and re-absorption of the x-rays within this enclosed cavity creates a quasi-black-body source which can, routinely, have an equivalent black-body temperature of 100–200 eV [61], depending on the cavity size. An exit aperture in the hohlraum allows radiation to fall onto a sample target which will often have an 'ablator' layer, consisting of a low Z material such as CH. The ablation of this layer leads to the generation of the pressure needed to drive a shock. In this way, Löwer *et al* [62] generated shocks of up to ~200 GPa (20 Mbar), in broad agreement with equation (2.24).

A clear advantage of this method for driving shocks is that non-uniformities in the optical laser profile are smoothed in the process of conversion to x-rays and thus a more uniform shock can be generated. On top of this, we can consider that the laser-plasma in which the x-rays are generated is physically separate from the shock sampled and this means that faster electrons created in the laser-plasma are far less likely to pre-heat the sample. A more recent use of this technique is described by Rothman *et al* [49], where highly accurate (1%) shock speed measurements were possible, allowing detailed equation of state studies for pressures up to ~200 GPa, using the impedance match technique described in section 2.6.

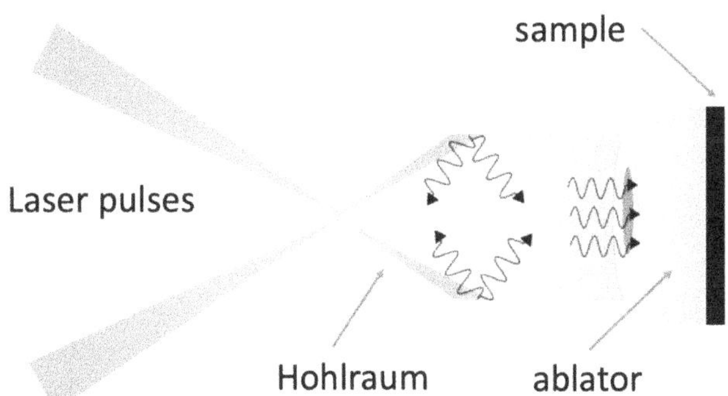

Figure 2.9. Schematic of a hohlraum target used to create an intense quasi-black-body radiation field. The peak of the emitted spectrum is typically about 0.5 keV, for which energy photons, the attenuation length is of order 1 μm in CH.

High power lasers are not the only source of intense x-rays for driving shocks. X-rays from Z-pinch facilities have also been used to indirectly drive Mbar pressure shocks [63]. In the Z-facility at the Sandia National Laboratory, a wire array pinch driven by a ~20 MA current can be used to generate up to 1.7 MJ of x-rays in bursts of a few nanoseconds duration [64]. The wire array is contained in a canister of typically ~2.5 cm in diameter and this containment creates a quasi-Planckian radiation field with an effective temperature of up to ~150 eV. By placing secondary hohlraum canisters around the central pinch, with typical diameters and lengths of 5–6 mm, such that they do not view the wire array directly, very uniform radiation fields of black-body temperatures 50–100 eV can be created to drive shock sample packages at the end of the secondary hohlraums.

For shocks driven by x-ray irradiation, it is import to take care that radiative pre-heat is minimised so as to have well defined initial conditions. For example, for an Au laser-plasma source, the conversion efficiency to M-band from laser energy can readily be of the order of a couple of percent [58] and the photon energy spans the 2–4 keV energy range. This means that for a low Z ablator layer, such as parylene, significant pre-heating of the shocked sample can be avoided only if the layer is >100 μm thick. For the work in [63], the irradiated ablator material is 100 μm of Al and the shock generated in this is either transmitted to an abutted Be sample material or used to accelerate the Al to form a flyer plate, as discussed below. For the hohlraum work in [49] an Au pre-heat shield, a few microns thick was used.

2.4 Ion beam driven shocks

A possible alternative to using lasers to drive shocks is to use ion beams. As we will see in the next chapter, ion beam facilities currently in development may provide volumetric heating of solid density matter sufficient to create WDM states directly. However, proposals have also been made to use this capability for the generation of shocks [65–71] and preliminary experiments have been carried out.

Amongst earlier experiments [65] Ewald *et al* have demonstrated the deposition of an average 1.5 MJ/kg using ~10^{11} U^{+28} ions in a microsecond pulse. The temperature reached was 0.2 eV and a shock wave of order 15 kbar (1.5 GPa) generated in a plexiglass sample abutted to the Pb absorber. In this experiment, the shock was driven in the direction of the ion beam. The Pb absorber was several mm thick, longer than the range of the ions. The shock pressure generated is not enough to bring us into the WDM regime. However, upgraded facilities such as the FAIR project [72], aim to deliver in excess of 10^{12} U^{+74} ions at over 1 GeV/u in a 50 ns pulse, where u is the atomic mass unit. This can be used to achieve energy deposition of up to 600 MJ/kg in Pb targets. The range of ions in matter, is more fully discussed in chapter 3 but we note that, for such high energy ions, the range would be over 1 cm, even for a high Z material like Pb [73], and uniform deposition of energy over a several mm path would be achieved. Combined with the possibility of sub-mm focal spots, this leads to other possible arrangements to drive shocks.

For example, consider the schematic in figure 2.10(a). Elliptical focusing of an ion beam to a spot with width as low as 100 μm is possible and indeed already reported

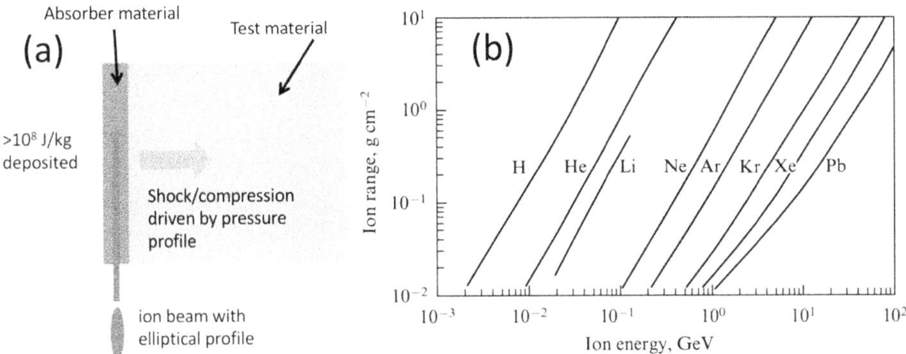

Figure 2.10. (a) Schematic of a potential experimental arrangement using an ion beam with a Bragg peak well beyond the target length. (b) Range of various ions in cold matter as a function of ion beam energy. If the range is in areal density, then it is similar for a wide range of target sample materials. Figure reproduced with permission from [73]. ©Uspekhi Fizicheskikh Nauk 2008.

for a 120 ns pulse of U^{+28} ions of 350 MeV/u in an experiment with uniform heating of a tungsten foil to temperatures of order 10^4 K [67]. From the data in figure 2.10(b), we can deduce that the range of such ions in Pb is over 6 mm and so we can see that, if we have a target width well within this range, we can uniformly heat an absorber material to high temperature at solid density, generating a pressure high enough to drive a strong shock into the sample material in a direction perpendicular to the ion beam direction. If we consider that the pressure generated is essentially the deposited energy density, then we can see that shocks in excess of 1 TPa can be generated. For a circular ion beam, this shock would be radial, although it would be 1-D. For elliptical profiles, we would be closer to a 1-D planar shock.

There are some key advantages to the use of ion beams in this manner. Firstly, we can note that deposition of 600 MJ/kg in Pb at solid density means the temperature is still under 100 eV. This means that there is no source of keV x-rays adjacent to the sample material and pre-heating is mitigated strongly. Secondly, as we noted above, for laser pulses with several nanoseconds duration, we have an issue with expansion of the high temperature coronal plasma. For ion beam experiments such as figure 2.10, this will not be such an issue and the use of pulses 10s nanoseconds means we can drive steady shocks for longer. Of course, for this timescale, depending on the target geometry, the hydrodynamic motion of the heated part of the target may start to become important and will need to be accounted for in modelling experiments. The ability to tailor the temporal behaviour of the ions beams over 10s of nanoseconds timescales means that there are key opportunities to study ramp compression where the entropy is kept low and off-Hugoniot states are generated. Grinenko *et al* [68] have discussed this possibility. The temperatures of up to 100 eV reached means that the ion beam heated material itself constitutes a volumetrically heated WDM state and we will return to the discussion of ion beam heating in chapter 3.

2.5 Flyer plate methods

Although, lasers can be used to drive shocks of pressures well above the Mbar (100 GPa) level, there are, as we have seen, some drawbacks. These are mainly connected to the pre-heating due to fast electrons and x-rays. A technique that avoids these issues is to use a flyer plate. In this method, a flat plate is accelerated to several km per second before colliding with the sample target, where pressures of several Mbar can be created. The collision generates a powerful shock wave without the pre-heating. In the next sub-sections, we will discuss several ways of accelerating a flyer plate and, following that, we will discuss the transmission of shocks from the plate to the sample.

2.5.1 Gas guns

In figure 2.11 we see a schematic (not to scale) of a typical two-stage light gas gun design [74]. In the first chamber, a charge of gunpowder is detonated. The rapidly expanding gases from the explosion drive a piston down the first stage barrel. So far, this is similar to a conventional rifle. The piston speed is limited by several factors, including the mass of the piston and amount of charge used. However, a key limitation is also the speed of sound in the expanding gases. For gunpowder gases, this is similar to the speed of sound in air (330 ms^{-1} under standard temperature and pressure, STP). Ahead of the projectile is a light gas, usually hydrogen, which has a sound speed at STP of 1270 ms^{-1}. As can be seen in the figure, the first section tube narrows towards the end where there is a diaphragm, which contains the light gas within this section. The piston, which is usually made of deformable material such as a plastic, is squeezed into the tapered section, raising the gas pressure, which bursts the diaphragm (typically mylar). The light gas can then expand rapidly behind the projectile, which can be in the form of a flyer plate. The high sound speed of the light gas allows for a much faster expansion and the flyer plate moves down the second stage tube at speeds of up to 8 km s^{-1}, whereas a single stage gas gun is limited to around 2.5 km s^{-1}. As the plate impacts a sample target, it can generate a strong

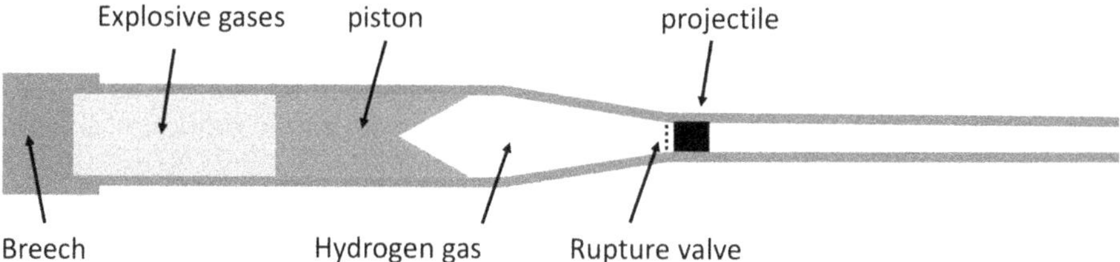

Figure 2.11. Schematic of a two-stage light gas gun. The first stage is similar to the breech of a gun and indeed rifles have been used for the first stage of small systems. Typically, the projectile is fired into a vacuum chamber (not shown) in which a target structure is placed to be struck by the projectile. A key restriction of such facilities is the limited shot rate coming from the use of explosives and the wear on the components. The JASPER (Joint Actinide Shock Physics Experimental Research) facility at LLNL for example, fires about 12 times a year.

shock. For example, shocks of up to 660 GPa have been generated in platinum with such a facility [75].

The launch tube diameter can be of order several cm with impactor thicknesses of a few mm. A key experimental constraint is the degree of 'flatness' in the impactor and target plates. The surfaces are typically lapped to a flatness of 1–2 μm with a similar degree of accuracy for the parallelism of the front and rear surfaces of the target. Typically flyer plates can be accelerated with tilts of less than 2 mrad. Across a 28 mm flyer plate this corresponds to a difference of only about 50 μm in the edge positions in the direction of travel. At 8 km s^{-1} this means a timing difference of 6 ns across the shock region.

By use of flash radiography, the impactor velocity has been measured to an accuracy of 0.1% [76] and the shock velocity to an accuracy of order 1%. The shock speed is measured in some cases by the insertion of electrical pins at different depths within the target. The pins are electrically isolated from the target but are butted up against a mylar sheet, typically a few microns thick. A bias voltage is applied to them. As the shock reaches the mylar, the compression creates an electrically conducting state. Since the pins and target are both grounded to the target chamber, a fast rising current flows, which is recorded. A sub-nanosecond temporal resolution is possible with this arrangement and this is short compared to the timescales of typically greater than 100 ns for the shock trajectory. Analysis of the shock rise times for an array of pins in the target can then be used to correct for any tilt in the impactor.

The development of a so called three-stage gas gun has been reported [77]. This is essentially an extension of the two-stage system, where the flyer plate is no longer a single material but has multiple layers to generate a graded density impactor. This strikes an intermediate buffer layer of polymethyl pentene that is abutted to the final flyer plate. The graded density leads to more gentle compression of the buffer layer, with less energy being dissipated as shock heating. The enhanced efficiency of the pressure transfer allows the final flyer plate to be accelerated to 10 km s^{-1}.

2.5.2 Magnetically driven flyer plates

The gas-guns discussed above can create conditions relevant to WDM studies, using flyer plates accelerated up to around 8 km s^{-1}. However, magnetically driven flyer plates can reach over 30 km s^{-1} [78, 79] allowing access to more extreme conditions. If we look at figure 2.12(a), we can see a simplified sketch of a circuit that, with a suitable pulsed current can generate a Lorentz force that accelerates a flyer plate. The circuit consists of a cathode shorted with an anode. The anode itself consists of a frame with a central panel, typically less than 1 mm thick, that forms the flyer plate.

The gap between the cathode and anode is around a millimetre wide and forms a current loop that, for a suitable, powerful pulsed current source, generates a large magnetic field. The magnetic pressure causes a stress wave that constricts both the anode and cathode. As described by Lemke *et al* [78], when this stress wave is released at the surface of the flyer that faces the target, the flyer then moves independently of the rest of the anode towards the target. This can happen because the magnetic force is sufficient to overcome the material strength.

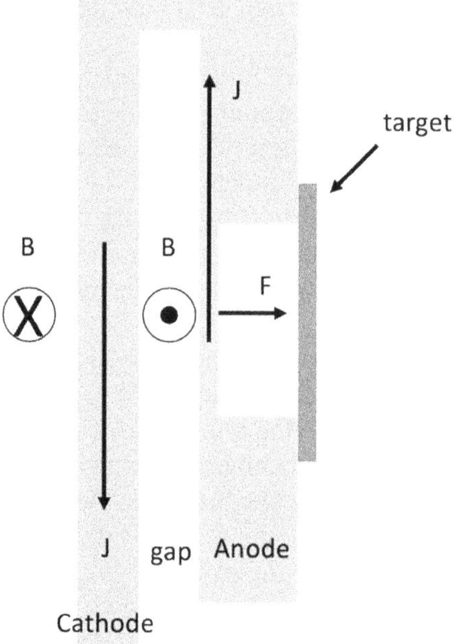

Figure 2.12. Simplified schematic of a magnetic flyer plate. The current flow and magnetic field directions are indicated as is the direction of the magnetic force accelerating the flyer plate portion of the anode.

As an example, instead of being used to generate x-ray drive, as described above, the Sandia Z machine [78], can generate the necessary B-field with a current of 20 MA rising linearly over about 200 ns. With a single flyer-plate anode and a stainless steel cathode, this has been used to create a field of order 1000 T. The magnetic pressure given by;

$$P_B = \frac{B^2}{2\mu_0} \tag{2.25}$$

is around 400 GPa, which is well in excess of the ~0.3 GPa yield strength of Al. As with the gas-gun, the flyer plates are of a few cm in dimension though generally only around a millimetre or less in thickness, a flight gap of ~5 mm to the target samples. Generally, 2-D MHD simulations that include inductance effects are needed to model the flyer plate velocity achieved, but simulations are in good agreement with experimental measurements made with VISAR diagnostics. In fact, the design used at Sandia allows up to four flyer plates placed around a central square cross-section cathode. This lowers the maximum magnetic pressure but still allows flyer-plate velocities of around twice that achievable in a two-stage gas gun.

2.5.3 Explosively driven shocks and compression

Early shock experiments used explosives and, even in more recent times, they are still used as high pressures over large scale targets are possible, driven by the use of up to

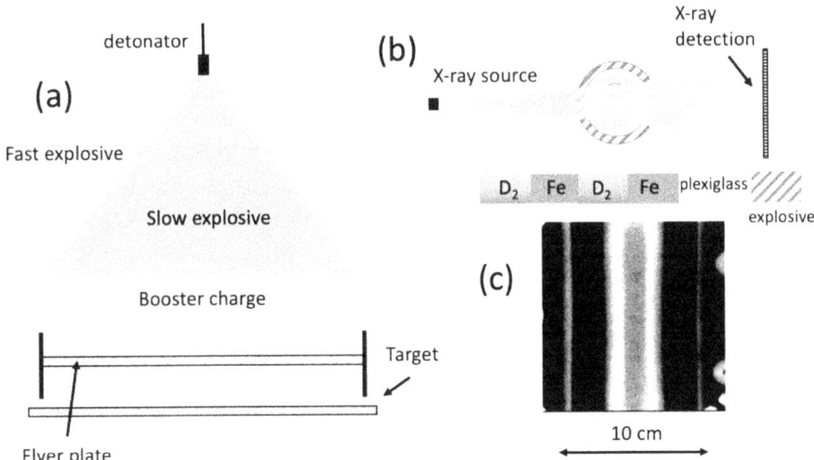

Figure 2.13. (a) A combination of fast detonation and slower detonation explosives has been used often to create and shock lens has been a widely used to create planar shocks. (b) Cylindrical compression of concentric cylinders of Fe filled with hydrogen or deuterium have been used to create far higher compression than available in planar compression (c) Typical radiograph of D_2, quasi-isentropically compressed in a cylindrical compression, reprinted with permission from [80] ©2007 by the American Physical Society.

several 10s of kg of explosive material [80]. A typical planar shock experimental arrangement is shown in sketch form in figure 2.13(a). As we can see, the first part of the charge is shaped to allow a planar shock to be generated. The conical charge has an outer layer made from an explosive with a fast detonation wave speed, whilst the inner part has a slower detonation wave. The detonation is initiated at the apex and proceeds along the outside rapidly. The middle, directly underneath the initial detonation is furthest from the exit plane but is detonated earliest and by the time the detonation wave reaches the outer part of the cone along the fast explosive, the wave-front is planar across the cone. This, of course, requires careful matching of the detonation speeds and shaping of the charges.

As in the figure, a typical WDM experiment would not use the shock directly by abutting the explosives to the sample (although this can, and has been, done for other purposes). Instead, a flyer plate is accelerated by the exiting explosion gases. In order to reduce deformation of the flyer plate, a gap is often used between the flyer plate and the explosive. This drives the plate in a gentler way. Nevertheless, flyer-plate velocities as high as 8 km s^{-1} have been achieved and this is sufficient to drive >100 GPa shocks in a sample.

Explosives are not just used for planar shocks. Both cylindrical and spherical shocks have been driven, in particular on experiments on hydrogen. For example, Fortov *et al* [80] have reported using a cylindrical arrangement of layers, as depicted in figures 2.13(b) and (c). As we see, the experimental package consists of concentric cylinders of Fe into which the H_2 or D_2 sample is filled. An outer layer of plastic separates the outer iron layer from the explosive and this helps to smooth non-uniformities in the shock before it reaches the iron cylinders. In their experiment, the shock driven into the target assembly reverberated between the iron and hydrogen layers, allowing a quasi-isentropic compression of D_2 to 300 GPa at a density of a little over 2 g cm^{-3}. The density was determined in this experiment by flash radiography of the sample assembly using an x-ray source generated with an MeV electron beam. The duration of the x-rays was ~300 ns, which is short compared to the probe time, which was approximately 40 μs after detonation. An additional diagnostic in such cylindrical experiments is the conductivity of the sample. As described in [81], two probes can be inserted from each end along the axis of the cylinder with a fixed gap between their end. An applied dc voltage will generate a current that depends on the dc conductivity of the sample under compression.

A key outcome of this and associated experiments is to observe evidence for the plasma phase transition that has been postulated for more than 50 years [82]. This is manifested as a change from an insulator phase to a metallic phase and a sudden jump in conductivity. The conductivity was seen by Fortov *et al* [81] to rise by 5 orders of magnitude on compression to ~2 g cm^{-3} with pressure above 100 GPa.

By extending the concentric implosion technique to the spherical case, quasi-isentropic compression of deuterium to pressures as high as 5500 GPa (55 Mbar) has been reported [83], using up to 54 kg of explosives and achieving densities of ~6 g cm^{-3}, more than 30 times the liquid density.

2.5.4 Laser driven flyer plates

It is worth mentioning that flyer plate driven shocks have been created with lasers as well as gas guns, though the latter is more common. A laser irradiated foil can be made to play the role of a flyer plate by either direct illumination with a laser beam or by being indirectly driven with an intense x-ray source. An example of the latter is presented in [84] where 25 kJ of laser light in a 1 ns pulse was focused into a gold hohlraum to create an intense x-ray drive that was incident on a 50 μm polystyrene ablator foil on top of a 3 μm Au flyer plate.

This flyer plate collided with a secondary stepped Au foil sample to generate a ~Gbar pressure shock wave, where the pressure was inferred from the speed of the shock. This, in turn, was measured by observing the optical emission from the rear of the Au steps as the shock emerged from the rear. As with the soft x-ray driven shock experiments discussed earlier, there is the potential advantage of reducing radiative and electron pre-heating. However, hydrodynamic instability of the flyer plate is an issue to contend with, as any non-uniformity or gradient in the irradiation of the flyer plate foil will lead to a non-flat impact and thus non-uniform shock drive. The small size of the plates accelerated and the issues of laser spot uniformity mean that, in general, this is a less used method of accelerating flyer plates.

2.6 Impedance matching

The technique of using an impactor or flyer plate to strike a target generally involves use of an important concept, known as *impedance matching*, where the impedance in question is given by;

$$Z = \rho_0 U \tag{2.26}$$

In subsonic work, U is the longitudinal sound speed, but, in cases of interest to us, it is the shock speed and ρ_0 is the initial density of the material in question. The concept applies for cases where we either have an impactor, as in flyer plates, or where the shock is transmitted across an existing interface, from one material to another, and we will look at both cases.

For flyer plates, the analysis of such experiments starts with the observation that across the interface between target and impactor, the normal stress and the particle velocity are continuous. If we take our analysis into the laboratory frame, we can consider the relationship between the impactor and target Hugoniots. If we look at figure 2.14(a), we can see the target Hugoniot plotted with the pressure increasing with particle velocity, as usual, whereas, for the impactor, the starting particle velocity is the velocity, U_i, to which it has been accelerated, just prior to impact, and the particle velocity behind the shock, in the laboratory frame, is given by $U_i - u_p^i$, where u_p^i is the particle velocity within the frame of the target, thus the Hugoniot, for the impactor, is shifted and reversed in direction.

An important point is made in figure 2.14(b). If the impactor is made from a reference material where the Hugoniot is known, then we only have to measure the impactor velocity to establish the Hugoniot curve in the diagram. If we then measure the shock speed in the target, we can use equation (2.2) to establish the solid line of slope $\rho_0 u_s$ (assuming $P_0 = 0$). The intercept of these two curves gives us the particle velocity in the target and thus, with both u_s and u_p derived for the unknown material, we can solve the Rankine–Hugoniot equations fully for the unknown target material.

In the special case that the target and impactor are made from the same material, then the requirement of continuous particle velocity across the interface means that the particle velocity is given by;

$$u_p = \frac{1}{2} U_i \tag{2.27}$$

An example of a standard material is, for example, fused silica (SiO_2), whose shock equation of state has been explored to over 1 TPa [85, 86] in both laser-driven and

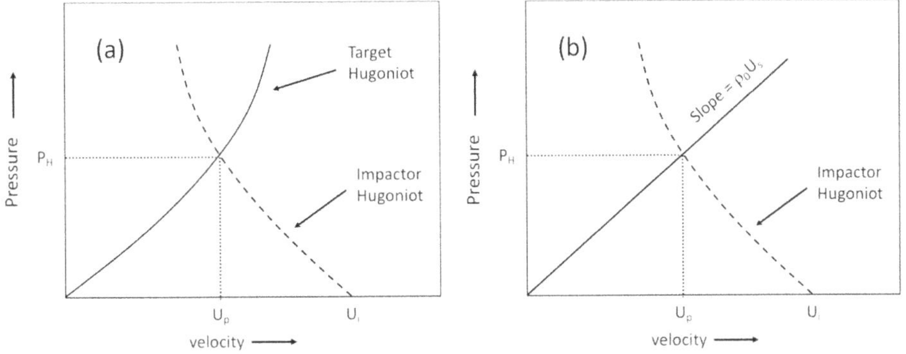

Figure 2.14. (a) As the impactor strikes the target, the particle velocity is continuous across the boundary and its value defines a point on the Hugoniot of both materials. (b) Assuming the shock velocity is measured in the target, and the impactor Hugoniot is known, we can establish a point on the unknown Hugoniot.

flyer plate drive shock experiments. Root *et al* [86], for example, have carried out extensive magnetically driven flyer plate experiments, where a VISAR system measured the flyer plate velocity through the fused silica sample and then the shock speed as the fused silica melted into a conducting fluid state under shock compression to 1100 GPa. In this way, a reference Hugoniot was accurately established.

Reflected shocks and release waves

In figure 2.15(a), we show a schematic of a fairly typical laser-driven shock impedance match experiment, with a multilayer target [87], although similar experiments can be carried out with flyer plates. A standard material is present, with a two-stepped layer, and measurement of the shock speed can be made by comparing the shock break-out times across the two steps. An unknown material is abutted to part of the standard sample and the shock speed through it can be measured by comparing shock break-out times with the stepped standard.

Knowledge of the shock speed for this standard material establishes the particle velocity at the standard/unknown material boundary. If the shock velocity is now observed for the unknown material, we will now have two of the five unknown quantities in the Rankine–Hugoniot equations and can establish a point on the shock Hugoniot. The unknown material can have either a higher or lower shock impedance and we can consider the outcome in either case, both of which are shown schematically in figure 2.15(b). If the unknown material has a higher impedance, a shock is reflected back into the standard material. This is essentially the same situation we discussed above with regard to using a low Z ablator layer to absorb the laser energy. The momentum change involved in reflecting the shock from the boundary enhances the pressure imparted to the unknown sample, as seen in figure 2.7. The standard material is reshocked onto a secondary Hugoniot that has as its starting condition the point on the principal Hugoniot to which the standard material is initially shocked. For the standard material, the new particle

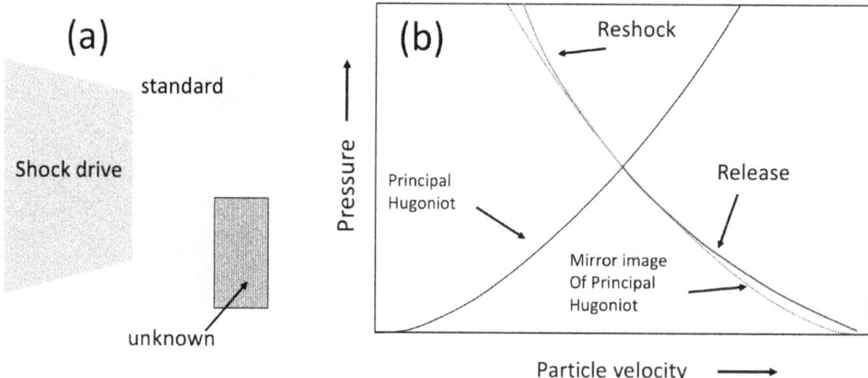

Figure 2.15. (a) Schematic of a typical laser-driven shock impedance match experiment with a standard whose Hugoniot is known and a sample whose Hugoniot is to be determined. (b) The reshock and release curves for the standard, depending on whether the impedance of the unknown sample is higher or lower.

velocity in the laboratory frame is given by the difference between the particle velocities for the initial and the new Hugoniots. With the help of equation (2.2), we can see that the new shock condition in the standard material is;

$$P^r - P^h = \rho^h\left(|u_p^r - u_p^h|u_s^r\right) \tag{2.28}$$

where the superscripts, h and r, refer to the states on the initial Hugoniot and the reflected shock Hugoniot, respectively.

On the other hand, if the unknown sample has a lower impedance, the wave reflection back into the standard is an isentropic release wave and the shock transmitted into the unknown material is a lower pressure than the initial shock pressure.

The particle velocity of the release wave in the standard material is now given by;

$$u = u_p^h - \int_{P^h}^{P^r} \frac{dP}{\rho^r c_s^r} \tag{2.29}$$

where the superscript, r, now refers to the release wave. The isentropic sound speed as a function of density, $c_s(\rho)$, needs to be known for the standard material in question. In order to get an accurate determination of the release isentrope, we need to know the equation of state for states on and off-Hugoniot. In the absence of very accurate equation of state data for off-Hugoniot states, a common approximation is that the reshock and release curve for the standard material in figure 2.15(b) is in fact a mirror image of the principal Hugoniot. This is in fact found to be quite accurate up to shock pressure of around 200 GPa, or where the impedances of the standard and unknown materials are close [88]. A consequence of this symmetry is that, in the absence of strong effects of radiation loss and thermal conduction, for shock-break-out into a vacuum, the expansion speed of the surface is expected to be $\sim 2u_p$.

Consideration of reshock and release are not just relevant to experiments of the type shown in figure 2.15. As we shall see in chapter 5, when using optical diagnostics such as VISAR and streaked optical pyrometry, we often use a transparent window fitted to the rear of the target, which can have either a higher or lower impedance than the sample material. In the absence of this window, the sample expands to form to a low density vapour/plasma that is highly absorbing of the VISAR probe laser and whose optical properties can complicate the interpretation of optical emission and optical reflectivity experiments [89]. The window prevents this, however, to be used effectively, its own optical properties under shock compression need to be characterised.

2.7 Diamond anvil cells

We finish this chapter with a section that, at first sight, does not really fit the theme of ramp or shock compression. As noted in chapter 1, the conditions accessed by a DAC, although extreme, are generally only considered to be on the margins of WDM. However, advances are being made and a DAC can create pressures and temperatures that are relevant to work on the Earth's interior and thus of interest for many whose interests span planetary sciences and WDM.

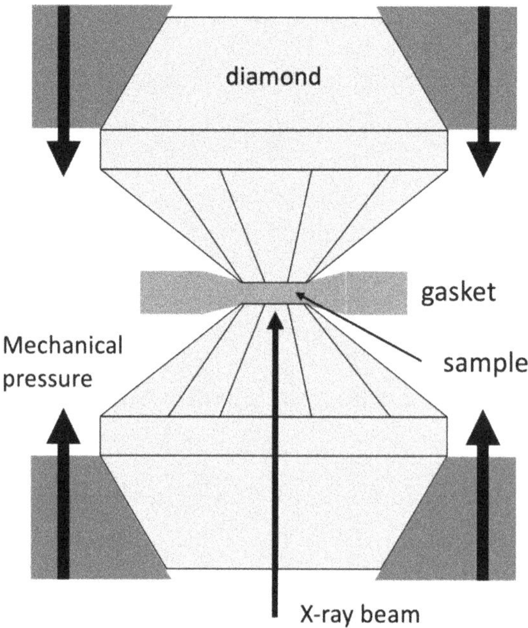

Figure 2.16. Simplified schematic of a diamond anvil cell. In this case, we show an x-ray beam that would, in many experiments, be used to measure diffraction from the sample.

In figure 2.16 we see the basic structure of a DAC [90–92]. A sample is placed between two gem quality diamonds whose flat culet faces are typically sub-mm in diameter. The large ratio between the size of the diamond, where the mechanical force is applied, and the culet allows a high pressure that can range up to around 700 GPa. For very high pressure work, the culet might be only of order a couple of hundred microns across and strong pressure gradients can be present. The sample is usually constrained by a deformable gasket which is typically made of rhenium, tungsten, or stainless steel. The volume sealed within the diamonds and gasket not only contains the sample but a pressure transmitting fluid, which can often be an inert gas such as helium. For lower temperature work, it is also common to include a small ruby, as the spectral position of laser fluorescence from ruby can be used as a pressure monitor. At temperatures relevant to WDM, the lines are broadened to the point they are not a viable diagnostic. However, they can be used to monitor the starting pressure for experiments where samples may be heated after compression. In order to create samples more relevant to WDM, heating is required, in addition to pressure. Whilst for some samples, resistive heating has been used, this is limited to temperatures a little below the desired regime. Laser heating can reach temperatures close to 6000 K with pressures above 300 GPa [93]. This is usually achieved with near IR lasers of order 50–200 W power. However, this is still at the low end of the WDM regime.

In order to use a DAC to access states more relevant to WDM research, in addition to heating, we can use a pre-compressed sample as a starting condition for shock compression, thus creating a new Hugoniot along which to sample the material, as described above in reference to the work of Eggert *et al* [16]. In that case, one of the diamonds was replaced with a flat diamond plate less than 500 μm thick. This allowed a high power laser to be focussed through the plate, onto the sample load, to allow a strong shock to be driven. The other diamond anvil was replaced by an anvil made from sapphire. This is because it has an optical transparency more suitable for measurement with a VISAR incident through the anvil, onto the rear of the sample, allowing the shock break-out to be recorded and both shock and particle velocity to be determined by impedance matching. Naturally, as with the windows mentioned above, it is important that the optical properties of the anvil, for both VISAR and optical emission measurements, are well understood, not just at ambient conditions but also under shock compression.

References

[1] Walsh J M and Christian R H 1955 *Phys. Rev.* **97** 1544–56

[2] Bancroft D, Peterson E L and Minshall S 1956 *J. Appl. Phys.* **27** 291

[3] Zel'dovich Y B and Razier Y P 1966 *Physics of Shock Waves and High-Temperature Hydrodynamic Phenomena* (New York: Academic)

[4] Landau L D and Lifshitz E M 1987 *Fluid Mechanics, Course of Theoretical Physics* 2nd edn 6 (Oxford: Pergamon)

[5] Al'tshuler L V 1965 *Sov. Phys. Usp.* **8** 52

[6] Al'tshuler L V, Krupnikov K K, Fortov V E and Funtikov A I 2004 *Her. Russ. Acad. Sci.* **74** 613–23

[7] Courant R and Friedrichs K O 1948 *Supersonic Flow and Shock Waves* (New York: Interscience)

[8] More R M, Warren K H, Young D A and Zimmerman G B 1988 *Phys. Fluids* **31** 3059–78

[9] Lyon S P and Johnson J D 1992 SESAME: The Los Alamos National Laboratory Equation of State Database *LANL Technical Report* LA-UR-92-3407

[10] Seigel A E 1977 *High Pressure Technology, II* ed I L Spain and J Paauwe (New York: Marcel Dekker)

[11] Denis G 2017 *Physics of Shock and Impact: Volume 1: Fundamentals and dynamic failure* (Bristol: IOP Publishing)

[12] Brown J M, Fritz J N and Dixon R S 2000 *J. Appl. Phys.* **88** 5496

[13] Eliezer S 2011 *Laser-Plasma Interactions and Applications: proceedings of the 68th Scottish Universities Summer Schools in Physics* (Berlin: Springer)

[14] Nellis W 2006 *AIP Conf. Proc.* **845** 115

[15] Ozaki N *et al* 2016 *Sci. Rep.* **6** 26000

[16] Eggert J H *et al* 2009 *AIP Conf. Proc* **1161** 26–31

[17] Jackel S, Salzmann D, Krumbein A and Eliezer S 1983 *Phys. Fluids* **26** 3138

[18] Coe S E, Willi O, Afshar-Rad T and Rose S J 1988 *Appl. Phys. Lett.* **53** 2383

[19] Riley D, Willi O, Rose S J and Afshar-Rad T 1989 *Europhys. Lett.* **10** 135

[20] Larsen J T and Lane S M 1994 *J. Quant. Spectrosc. Radiat. Transfer* **51** 179

[21] Brunton G, Erbert G, Browning D and Tse E 2012 *Fusion Eng. Des.* **87** 1940

[22] Kidder R E 1979 *Nucl. Fusion* **19** 223

[23] Nuckolls J M, Wood L, Thiessen A and Zimmerman G 1972 *Nature* **239** 139

[24] Wang J *et al* 2013 *J. Appl. Phys.* **114** 023513
[25] Koenig M *et al* 2010 *High Energy Density Phys.* **6** 210
[26] Amadou N *et al* 2015 *Phys. Plasmas* **22** 022705
[27] Mei Q S and Lu K 2007 *Prog. Mater. Sci.* **52** 1175–262
[28] Anzellini S, Dewaele A, Mezouar M, Loubeyre P and Morard G 2013 *Sci.* **340** 464–6
[29] Yoo C S, Holmes N C, Ross M, Webb D J and Pike C 1993 *Phys. Rev. Lett.* **70** 3931–4
[30] Ahrens T J, Holland K G and Chen G Q 1998 *Shock Compression of Condensed Matter 1997* ed S C Schmidt *et al* (Woodbury, NY: AIP Press) pp 133–6
[31] Nguyen Jeffrey H and Holmes Neil C 2004 *Nature* **427** 339–42
[32] Dolan D H and Gupta Y M 1978 *Phys. Rev. Lett.* **40** 1391
[33] Turneaure S J, Sharma S M and Gupta Y M 2018 *Phys. Rev. Lett.* **121** 135701
[34] Seagle C T, Desjarlais M P, Porwitzky A J and Jensen B J 2020 *Phys. Rev.* B **102** 054102
[35] Trainor R J, Shaner J W, Auerbach J M and Holmes N C 1979 *Phys. Rev. Lett.* **42** 1154
[36] Veeser L R and Solem J C 2004 *J. Chem. Phys.* **121** 9050–7
[37] Max C E 1982 *Physics of Laser Fusion Vol 1: Theory of the Coronal Plasma in Laser-Fusion Targets* UCRL-53107
[38] Atzeni S and Meyer ter Vehn J 2004 *The Physics of Inertial Fusion* (Oxford: Clarendon)
[39] Kruer W L 2003 *The Physics Of Laser Plasma Interactions* (Boulder, CO: Westview Press)
[40] Garban-Labaune C, Fabre E, Max C E, Fabbro R, Amiranoff F, Virmont J, Weinfeld M and Michard A 1982 *Phys. Rev. Lett.* **48** 1018–21
[41] Campbell E M 1992 *Phys. Fluids* B **4** 3781
[42] Trainor R J and Lee Y T 1982 *Phys. Fluids* **25** 1898
[43] Yaakobi B and Bristow T C 1977 *Phys. Rev. Lett.* **38** 350
[44] Goldsack T J, Kilkenny J D and MacGowan B J 1982 *Phys. Fluids* **25** 1634
[45] Hauer A, Mead W C, Willi O, Kilkenny J D, Bradley D K, Tabatabaei S D and Hooker C 1984 *Phys. Rev. Lett.* **53** 2563–6
[46] Thompson P C, Roberts P D, Freeman N J and Flynn P T G 1981 *J. Phys.* D **14** 1215
[47] Szichman H and Eliezer S 1992 *Laser Part. Beams* **8** 73
[48] Henke B L, Gullikson E M and Davis J C 1993 *At. Data Nucl. Data Tables* **54** 181–342
[49] Rothman S D, Evans A M, Horsfield C J, Graham P and Thomas B R 2002 *Phys. Plasmas* **9** 1721–33
[50] Swift D C and Kraus R G 2007 *Phys. Rev.* E **77** 066402
[51] Ng A, Cottet F, DaSilva L, Chiu G and Piriz A R 1988 *Phys. Rev.* A **38** 5289–93
[52] Kato Y *et al* 1984 *Phys. Rev. Lett.* **53** 1057
[53] Pepler D A *et al* 1995 *Proc. SPIE* **2404** 258
[54] Lehmberg R and Obenschain S P 1983 *Opt. Commun.* **46** 27
[55] Obenschain S P *et al* 1986 *Phys. Rev. Lett.* **56** 2807–10
[56] Mochizuki T *et al* 1986 *Phys. Rev.* A **33** 525
[57] Goldstone P *et al* 1987 *Phys. Rev. Lett.* **59** 56
[58] Kania D R *et al* 1992 *Phys. Rev.* A **46** 7853
[59] Pakula R and Sigel R 1985 *Phys. Fluids* **28** 232–44
[60] Pakula R 1985 *Phys. Fluids* **29** 1340
[61] Decker C *et al* 1997 *Phys. Rev. Lett.* **79** 1491–4
[62] Löwer T *et al* 1994 *Phys. Rev. Lett.* **72** 3186
[63] Bailey J E *et al* 2000 *J. Quant. Spectrosc. Radiat. Transfer* **65** 31–42
[64] Spielman R B *et al* 2000 *Phys. Plasmas* **5** 2105

[65] Dewald E *et al* 2003 *IEEE Trans. Plasma Sci.* **31** 221–6
[66] Constantin C *et al* 2004 *Laser Part. Beams* **22** 59–63
[67] Ni P A *et al* 2008 *Laser Part. Beams* **26** 583–9
[68] Grinenko A, Gericke D O and Varentsov D 2009 *Laser Part. Beams* **27** 595–600
[69] Hoffmann D H H *et al* 2005 *Laser Part. Beams* **23** 47–53
[70] Bieniosek F M *et al* 2010 *J. Phys.: Conf. Ser.* **72** 032028
[71] Tauschwitz A *et al* 2008 *J. Phys.: Conf. Ser.* **112** 032074
[72] Tahir N A *et al* 2005 *Contrib. Plasma Phys.* **45** 229–35
[73] Fortov V E, Hoffmann D H H and Sharkov B Y 2008 *Phys.-Usp.* **51** 109–31
[74] Jones A H, Isbell W M and Maiden C J 1966 *J. Appl. Phys.* **37** 3493–9
[75] Holmes N C, Moriarty J A, Gathers G R and Nellis W J 1989 *J. Appl. Phys.* **66** 2962–7
[76] Mitchell A C and Nellis W J 1981 *J. Appl. Phys.* **52** 3363
[77] Wang X *et al* 2019 *Rev. Sci. Instrum.* **90** 013903
[78] Lemke R W *et al* 2005 *J. Appl. Phys.* **98** 073530
[79] Knudson M D *et al* 2003 *J. Appl. Phys.* **94** 4420–31
[80] Fortov V E *et al* 2007 *Phys. Rev. Lett.* **99** 185001
[81] Fortov V E and Mintsev V B 2005 *Plasma Phys. Control. Fusion* **47** A65–72
[82] Norman G E and Saitov I M 2019 *Contrib. Plasma Phys.* **59** e201800182
[83] Mochalov M A *et al* 2017 *J. Exp. Theor. Phys.* **124** 592–620
[84] Cauble R *et al* 1993 *Phys. Rev. Lett.* **70** 2102
[85] McCoy C A *et al* 2016 *J. Appl. Phys.* **119** 215901
[86] Root S, Townsend J P and Knudson M D 2019 *J. Appl. Phys.* **126** 165901
[87] Batani D *et al* 2002 *Phys. Rev. Lett.* **88** 235502
[88] Celliers P M, Collins G W, Hicks D G and Eggert J H 2005 *J. Appl. Phys.* **98** 113529
[89] Celliers P and Ng A 1993 *Phys. Rev.* E **47** 3547–65
[90] Ming L and Basset W 1974 *Rev. Sci. Instrum.* **45** 1115
[91] Anzellini S and Boccato S 2020 *Crystals* **10** 459
[92] Li B *et al* 2018 *Proc. Natl. Acad. Sci. USA* **115** 1713–7
[93] Tateno S, Hirose K, Ohishi Y and Tatsumi Y 2015 *Science* **330** 359–61

IOP Publishing

Warm Dense Matter
Laboratory generation and diagnosis
David Riley

Chapter 3

Volumetric heating of warm dense matter

3.1 X-ray heating

3.1.1 Laser-plasma sources for x-ray heating

The generation of x-rays by focusing intense laser pulses onto a target has been investigated for several decades now [1–8]. It is well known that significant conversion of laser energy to x-ray energy in the keV and sub-keV ranges can be achieved by irradiating mid to high Z samples. The x-rays come from three basic mechanisms, which are discussed in detail, for example, in [9, 10]. Firstly, there is the free–free emission that arises when the free electrons in the plasma collide with ions and are scattered into other free states to generate bremsstrahlung (braking radiation). For a Maxwellian electron distribution, the radiated power can be given by;

$$P_{\text{brem}}(\omega) = \frac{32}{3}\left(\frac{\pi}{3}\right)^{1/2} c r_0^2 \left(\frac{E_{\text{hy}}}{T_e}\right)^{1/2} \bar{Z}^2 n_i n_e \exp\left(-\frac{\hbar\omega}{T_e}\right) G_{ff}(\omega, T_e) \tag{3.1}$$

where $r_0 = e^2/m_e c^2$ is the electron radius, in cgs form, and $E_{\text{hy}} = 13.605\,\text{eV}$ is the hydrogen atom ground state binding energy. The factor $G_{ff}(\omega, T_e)$ is the so-called free–free *Gaunt factor* that accounts for quantum mechanical effects but is commonly set to equal unity [10, 11]. The units of $P_{\text{brem}}(\omega)$ are in energy emitted per second per cm^3 per unit energy in spectral range (for example, eV/cm^3/s/eV). This is a smooth emission spectrum, with an intensity that falls off with photon frequency. The intensity depends strongly on the density, as it is proportional to both the electron and ion densities. Since the laser can penetrate up to the critical electron density (see chapter 2), and this varies as $\sim 1/\lambda^2$, we can see that there should be a strong dependence on the incident laser wavelength. The emission also depends on average ionisation, $\bar{Z}$, and so will be strongest for higher Z materials.

The next source of x-rays to consider is *recombination radiation* or free-bound radiation, which arises when free electrons recombine into a bound state of the ions by losing their energy to a photon in the process of radiative recombination. This mechanism also gives rise to a continuous spectrum, but with a minimum photon energy that is equal to the binding energy of the state to which recombination occurs, adjusted to account for ionisation potential depression. This leads to a so-called *recombination edge*, below which photon energy, no free-bound emission is emitted. In this case, we can write the power emitted by recombination into a level i, with principal quantum number, n, for an initial ion stage z as;

$$P_{\text{bfree}}(\omega) = \frac{64}{3}\left(\frac{\pi}{3}\right)^{1/3} c r_0^2 N_{i,z} n_e [E_{i,z'}/T_e]^{3/2} \times (1/n^3)\exp(-\hbar\omega/Te_e)[1 - P_{i,z}]G_{bf}(\omega, T_e) \quad (3.2)$$

where z is the initial ion stage and $z' = z - 1$ is the recombined ion stage. Again, we see there is a Gaunt factor, that is often approximated to unity [11] and the binding energy of the recombined state, $E_{i,z'}$ is calculated including continuum lowering. If the occupation fraction of the initial ion stage level $P_{i,z}$ is unity, then recombination is blocked.

Although the above are two important processes in hot plasmas, our purpose here is to generate a source for radiative heating of warm dense matter (WDM) and, for us, the most important x-ray generation mechanism is line emission. This is the process which generates the most intense x-ray spectra in laser-plasmas. In chapter 4 we will explore laser-plasma sources suited to diagnostics, which will include those emitting line spectra with narrow bandwidths suitable for x-ray scattering, as well as those emitting broad, smooth spectra, suitable for absorption measurements. However, in this chapter, we are concerned with the uniform heating of a WDM sample and our main considerations are going to be focussed, less on the spectral features, and more on the efficiency of the source in a spectral region where we can expect good penetration of the photons into the sample. As we shall see, this generally means using mid to high Z elements where, for example, we have significant L-shell ($n \to 2$) and M-shell ($n \to 3$) emission spectra in the keV photon regime and there are a large number of transitions from each ion stage. For a given atomic transition from state i to state j, the spontaneous decay rate (Einstein A-coefficient) depends on the oscillator strength, f_{ij}, and the square of the transition energy, ΔE;

$$A_{ij} = \frac{2e^2}{(\hbar c)^2 m_e c}(\Delta E)^2 f_{ij} \quad (3.3)$$

The actual conversion efficiency of laser-energy to x-ray energy will depend, not only on the material, but on the irradiation conditions, especially focussed intensity, pulse duration and wavelength. For nanosecond pulses with sub-micron laser wavelength, high absorption at high plasma density is possible [12] and the conversion of laser energy to x-ray energy can be substantial for elements such as

Au. In this case, a large fraction of the emitted x-ray energy appears in the sub-keV quasi-continuum N-shell and O-shell bands ($n \rightarrow 4$ and 5, respectively) [4]. In these bands, the emission lines are very closely spaced and subject to broadening effects such as Doppler and Stark broadening, such that they merge into one another, forming an unresolved transition array (UTA). In figure 3.1, we can see a comparison of experimentally measured Au emission with a scaled black-body. The Au data is extracted from the tables of Kania *et al* [5], close to the peak of emission for a 1 ns pulse with 0.53 μm wavelength, with a peak focussed intensity of $\sim 3 \times 10^{14}$ W cm^{-2}. The conversion of laser energy to x-rays transmitted through the foil was around 6% in this case. As we can see, the spectral shape of the emission can be approximated by a quasi-black-body with an equivalent temperature of around 180 eV. The intensity of the emission is, however, lower than for an actual black-body at this temperature and the black-body curve has been scaled by 50% to better show the spectral comparison. In cases such as this, where the spectral shape is given by a black-body but the emission intensity is lower, we would generally say that the emission is represented by a 'grey-body'.

Both the spectral shape and actual emission intensity are important in determining the heating that such a source can provide to a sample. For this reason, it is common, in simulations, to represent the heating radiation by an effective spectral temperature along with a dilution factor. An alternative way to look at this is that there is an equivalent black-body emission temperature and a separate spectral temperature. For this case, where the dilution of intensity is a factor of two, the well-known scaling of total emission with temperature for a black body;

$$I = \sigma_{SB} T^4 \tag{3.4}$$

where, in SI units, the Stefan–Boltzmann constant has a value, $\sigma_{SB} = 5.67 \times 10^{-8}$ Wm^{-2}K^{-4}, means that we can derive an emission temperature of around 150 eV. For this temperature, the flux of x-rays from the surface, in units typical in the

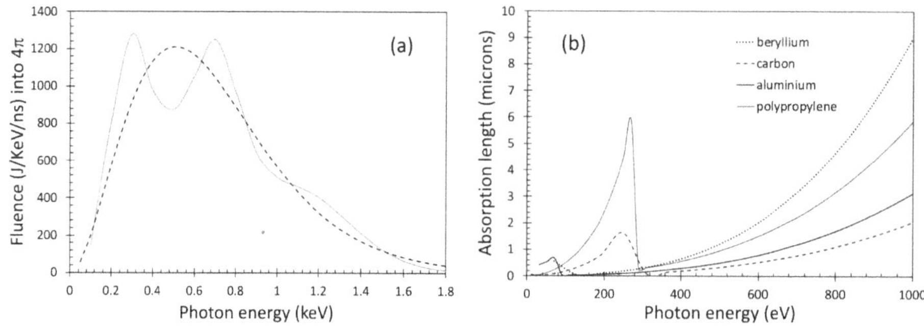

Figure 3.1. (a) Experimental emission spectrum from a gold foil target, with data extracted from [5]. The peaks around 300eV and 800 eV are the O-shell and N-shell bands, respectively. The solid line is the data and the dashed line is for a black body at a temperature of 180 eV, where we have scaled the black-body intensity down by a factor of 2. For an enclosed cavity, the spectral and emission temperatures can readily reach over 200 eV [13] and, in the largest laser facilities, may reach over 300 eV [14]. (b) Mean free path for absorption of sub-keV photons in a selection of materials [15].

literature, is 5×10^{13} W cm^{-2}. The detailed transport of radiation in a plasma is a complex topic that is dealt with in many textbooks and reviews, for example [16, 17]. We do not need to explore it in detail here, because we are interested in the volumetric heating of WDM, using photons that can penetrate the sample easily, and so we focus more on the photon mean free path for the cold sample. In figure 3.1(b), we show the mean free path as a function of photon energy for a series of low to mid-Z materials [15]. As we can see, even though the flux for a typical quasi-black-body drive is impressive, at the typical photon energies we would find volumetric heating difficult, except for the lowest Z elements, or in cases where at least one dimension of a sample is only microns in extent.

Good uniformity of heating, in fact, requires a photon absorption length (or photon mean free path) greater than the typical sample dimension. However, this implies a low efficiency of energy deposition and, for this reason, a high available x-ray flux is a necessity. Because of the need for a long photon mean free path, we generally require a source in the harder (>1 keV) x-ray regime. For the case of higher Z materials, such as Au discussed above, we can consider the M-shell emission. As shown in figure 3.2(a), this can span more than a keV in the few keV regime. Kania *et al* [5] have reported up to 2% conversion into M-band x-rays transmitted through a laser irradiated Au foil. However, as reported by Dewald *et al* [18], at intensities of $\sim 10^{15}$ W cm^{-2}, the conversion efficiency into the M-band of Au (2–5 keV) can reach up to 20% in a hohlraum target, with total conversion of laser energy to x-rays of over 70%. With above 13 kJ of input laser energy at 351 nm, this implies over 250 J/sr of M-shell radiation can be generated in a 1 ns pulse. This is a significant improvement over a flat foil, but we should also consider the geometry of any radiative heating experiment, where access for diagnostics also needs to be considered. With a laser-drive foil source, we may be able to place the WDM sample closer to the source and thus achieve comparable or better heating.

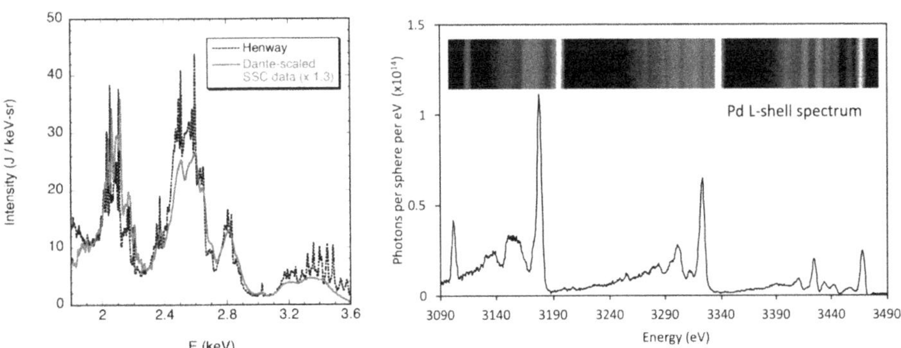

Figure 3.2. (a) M-band emission from a Au plasma irradiated at 351 nm with ~2.3 ns pulses focussed into a hohlraum target, reproduced from Robey *et al* [52], with the permission of AIP Publishing ©AIP 2005. (b) Experimental L-shell emission spectrum from a 50 nm Pd foil irradiated with 527 nm wavelength laser beams with 200 ps FWHM duration. Intensity on target was $\sim 10^{15}$ W cm^{-2} and conversion to L-shell x-rays was estimated at 4%. Similar data from the same experiment can be seen in [20]. The emission seen comes from Ne-like to Al-like ion stages.

As an alternative, for mid-Z materials, with roughly $Z = 40$–50, we can create L-band spectra in the few keV range. An example spectrum for Pd ($Z = 46$) is shown in figure 3.2(b). For laser irradiated foils, we can reach several percent conversion of laser light to L-shell x-rays depending on the pulse duration and focusing conditions. Phillion and Hailey [19] report conversion of around 2% of laser energy into x-rays over 4π sr, for 527nm laser pulses of 120 ps duration, whilst Kettle *et al* [20] report 4% conversion for 200 ps pulses at the same laser wavelength and similar intensity of $\sim 10^{15}$ W cm^{-2}. Hu *et al* [21] find conversion efficiencies at a similar level for Ag, Pd and Mo targets irradiated with 2 ns duration pulses, also at 527nm. We can reach higher conversions by the use of different target types. For example, Back *et al* [22] have used gas-bag targets filled with Xe and illuminated with 2 ns duration pulses of 0.351 nm laser light to reach L-shell conversion efficiency of around 10% for the 4–7 keV photon range. This is possible since, compared to foil targets, much less energy is used in hydrodynamic expansion and emission of broad-band, sub-keV photons.

A good example of the use of L-shell radiation for volumetric heating can be seen in the first experiment to demonstrate x-ray Thomson scattering (see chapter 4) [23]. In that experiment, a cylindrical sub-millimetre scale Be target was heated by L-band x-rays in the 2.7–3.4 keV range originating from an outer shell of Rh coated around the outside of the sample. The mean free path of these photons in Be ranges from 180 to 380 μm. Combined with radial symmetry, this allowed a relatively uniform electron density of over 3×10^{23} cm^{-3} at a temperature of 53 eV to be created. For this experiment, the ion–ion coupling parameter was of order unity. Furthermore, the Fermi energy of the electrons at this density is approximately 15 eV and thus $E_F/kT \sim 0.3$. This illustrates that both strong coupling and partial degeneracy, two of the key features of WDM, were present in the sample created.

More recently, Kettle *et al* [24] used Pd L-shell emission, from 3 to 3.5 keV, to volumetrically heat Al to WDM conditions of ~1 eV at solid density in order to probe free–free opacity in the XUV radiation at wavelengths longer than 17.1 nm (the L-edge for Al). In this experiment, the need to probe with XUV meant that the foils were of 200–800 nm thickness and, since the mean free path of the heating photons was 5–7 μm, with double sided heating, uniformity of better than 10% was achieved. This meant that, even for the thickest foils, only 11%–15% of the x-rays are absorbed. However, since there is such low absorption, it is more or less linear with thickness and the energy per atom absorbed was similar across the thickness range used. One consequence of using such a thin foil, in order to allow XUV probing, is that rapid decompression is to be expected and the experiment must account for this. In that particular experiment, a 200 ps optical pulse was used to create a source L-shell x-rays that rose to a peak over a time of about 150 ps. Hydrodynamic simulation for a 300 nm foil showed that the heated foil was at a temperature of around 0.8 eV after 100 ps, when the density of bulk of the foil was still close to solid. We can understand this by considering that the pressure at this point was about 25 GPa and, if we take the sound speed to be as given by equation (1.48) and assume the adiabatic index, $\gamma \sim 1$, we can estimate a sound speed of about 3 km s^{-1}. For this thickness of foil, this would indicate a decompression time of order 100 ps. In this case, a rapid heating time was necessary and the probing, with sub-ps high harmonic radiation, needed to occur before the peak heating, if close to solid density was to be preserved. The ion–ion coupling achieved in this case was of order $\Gamma\sim$ 50–80 with $E_F/kT\sim$ 0.1–0.15, and the fractional ionisation expected was $\bar{Z}/Z$ ~0.2, again showing the presence of key WDM features.

3.1.2 Experimental considerations: uniformity of heating

In the discussion above, we have noted two experiments that use volumetric heating to create a sample. In both cases, the uniformity of the sample, in the direction of travel for the photons through the sample, is affected by their absorption length and thus the spectrum of the source has an impact. However, the uniformity in other directions is also affected by the geometry of the heating. In the case of Glenzer *et al* [23], multiple beams are used to get close to cylindrical symmetry, but with the source adjacent to the Be sample, to allow for strong coupling of the heating radiation with the sample. In Kettle *et al* [24], fewer beams were available and access for probing the sample, without interference from the heating source was required. To accommodate this, the sources were offset from the sample. This has the disadvantage of diluting the radiation field before it arrives at the sample. However, a key advantage is that this can facilitate probing and emission diagnostics. Another advantage is that it also allows spectral filtering of the source, especially to remove low energy photons that have a short absorption length and would thus contribute to non-uniformity of bulk sample heating.

In such cases, we can estimate the level of heating that is possible for a separate source and WDM sample readily enough if we have a source that is well characterised. Let us make the approximation that the laser generating the source is focussed into a circular flat top distribution, with radius R; which is not unreasonable, given development in random phase plate technology. We can divide the focal spot region into annuli of area $2\pi r\mathrm{d}r$ ($0 \leqslant r \leqslant R$) and assume each has a radiance of σ_x W cm^{-2} sr^{-1}. Taking a point, z, along the central axis, normal to the plane of the source, the distance, from a point on the annulus, is given by $s^2 = r^2 + z^2$ and the angle of view is given by $\cos\theta = z/s$, as seen in figure 3.3. The plane of the heated sample (P in the figure) is also at the same angle to the x-rays coming from the source. This means that the flux crossing the sample plane, contributed from each annulus, can be integrated to give;

$$F_x(\mathrm{W\ cm^{-2}}) = 2\pi\sigma_x \int_0^R \frac{r}{(r^2 + z^2)}\mathrm{d}r \tag{3.5}$$

With standard integrals, it is straightforward to evaluate the heating as a function of distance. For the flat-topped source case, we can fit the on-axis flux as a function of distance by a scaling of the form;

$$F_x(\mathrm{W\ cm^{-2}}) = \frac{A_0}{(R^2 + z^2)} \tag{3.6}$$

where $z = 0$ gives us the power emitted per unit area from the sample surface. In many cases, the focal spot will not create a flat-topped source and we can use a

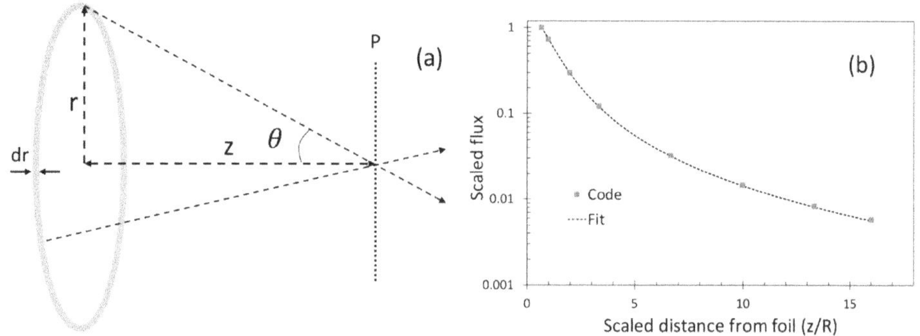

Figure 3.3. (a) Schematic geometry for heating with a separate source. (b) Relative heating flux, as a function of distance along z, in units of the source radius, for a flat-topped emission distribution.

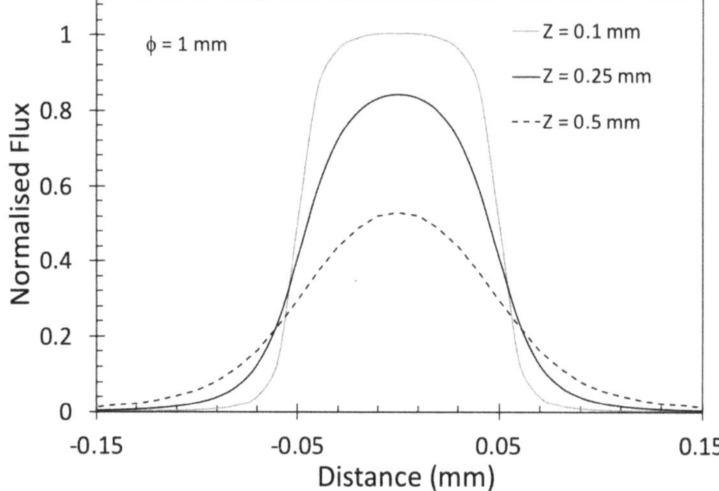

Figure 3.4. Simulation of lateral flux profile for a plane parallel to a 1 mm flat-topped source focal spot. At distances small compared to the spot size we see a roughly flat-topped flux but this rapidly changes as we move further away.

simple computer program to calculate the flux with equation (3.5) modified to account for radial variation in σ_X. We can also use a computer program to calculate the flux in a lateral direction parallel to the source. An example, for a flat-topped radial source, is shown in figure 3.4. In this figure, the width of the flux distribution depends on the ratio of the distance along z and the radius of the focal spot and we can estimate, for example, that with a focal spot diameter of 200 μm and an offset of 1.0 mm, the FHWM of the distribution is roughly 1.5 mm and this sets a limit to the area over which we can achieve uniform heating.

For the case of Pd L-shell emission irradiating a thin Al foil, discussed above, the peak radiance of the source was $\sigma_X \sim 10^{13}$ W cm^{-2} sr^{-1} with a ~100 μm radius focal spot. With angular behaviour of the emission folded in (cosine of angle from normal assumed), we can use equation (3.5) to calculate an on-axis flux of around 3.5×10^{11} W cm^{-2} at 1 mm distance. In this case, the absorption of ~3 keV photons is weak but double sided irradiance led to approximately 17 eV/atom total absorbed energy.

An important consideration, in any volumetric heating experiment, is the effect of softer x-rays, which can have a much shorter absorption length. This results in non-uniform heating, as the outer part of the sample can absorb significant additional energy from lower energy photons. This problem is often mitigated by use of a filter layer. Typically, a material such as CH is employed, as it is quite transparent to the desired keV heating x-rays whilst significantly blocking the softer x-rays. In the Kettle *et al* example, given above, the heated Pd layer produced a broad quasi-black body continuum with an effective spectral temperature of ~170 eV (figure 3.5). This means that there is a peak of emission at around 0.5 keV. This was removed by the

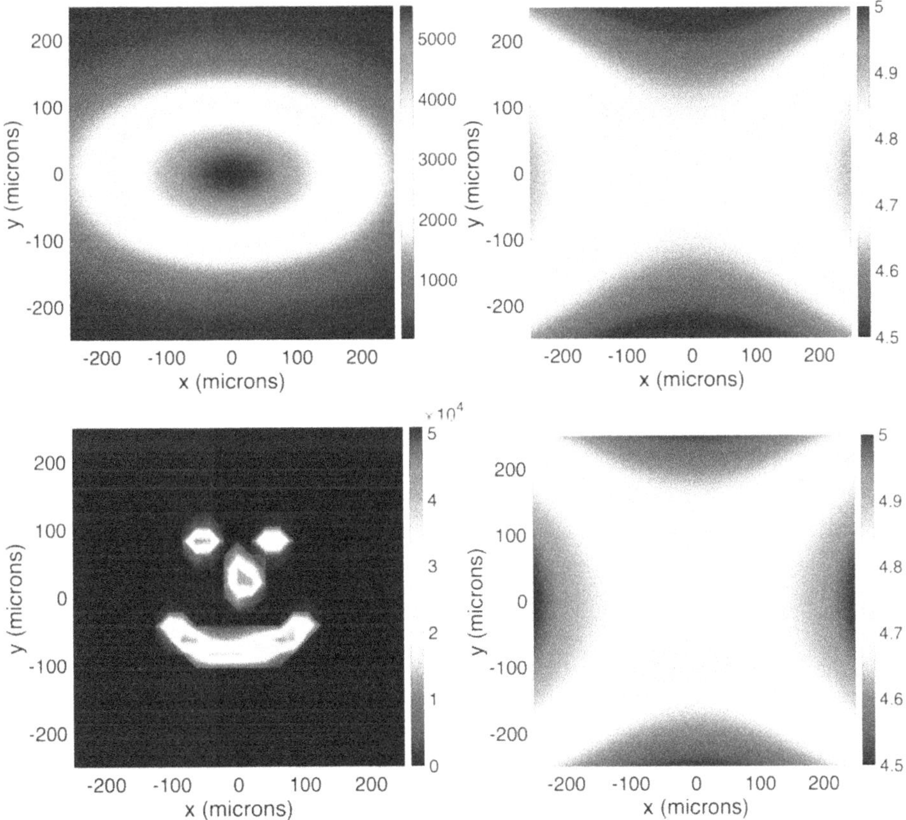

Figure 3.5. Simulation of the uniformity of heating for an Al foil between two parallel Pd foils set 2 mm apart with the sample foil at 45° as described in Kettle *et al* [20]. The two images on the left are the assumed distribution of x-ray emission from the Pd foil whilst, on the right, the images show the resultant distribution on the target foil, placed 1mm away at an angle of 45°. As we see, the uniformity of heating is not likely to be very sensitive to the distribution of the laser energy in the focal spot. This is a key advantage of using a sample foil offset from the heating foil. Simulations courtesy of C Hyland.

use of a 20 μm thick CH substrate onto which the Pd layer was coated, and through which, the L-shell x-rays could penetrate whilst the softer broad-band emission was attenuated by a factor >1000. In figure 3.5, we can see simulations of the spatial uniformity across the target for a typically expected smooth source. We can also see what happens when we assume a non-uniform (and in this case somewhat whimsical) source. What we see is that the uniformity on target is not really affected, a significant advantage of a separate source.

In summary, for radiative heating with laser-plasma sources, a key experimental issue is the question of whether the x-ray heating layer forms part of the sample target or whether, the need to have access for a particular diagnostic requires a separate target. In the case of Glenzer *et al* [23] the large, mm-scale, cylindrical geometry, and large number of beams available, allowed the heating layer to be coated directly onto the Be sample, with uniform heating and without compromising the use of a keV x-ray scattering probe. By contrast, the Kettle *et al* experiment required that an XUV probe be used and the x-ray emitting layer would, in this case, dominate the measurement, and so a separate heating source was chosen.

3.1.3 X-ray free electron sources

An important advancement of the last decade is the development of x-ray free electron lasers that have been applied to WDM experiments [25, 26]. It is worth noting that a key advantage of x-ray free electron lasers is that they produce an intense, highly collimated beam that is tuneable to over 10 keV and can be used for uniform volumetric heating, without the presence of an unwanted softer x-ray component, see for example Lévy *et al* [27]. The pulse durations are typically 10–100 fs. In the example of figure 3.6,

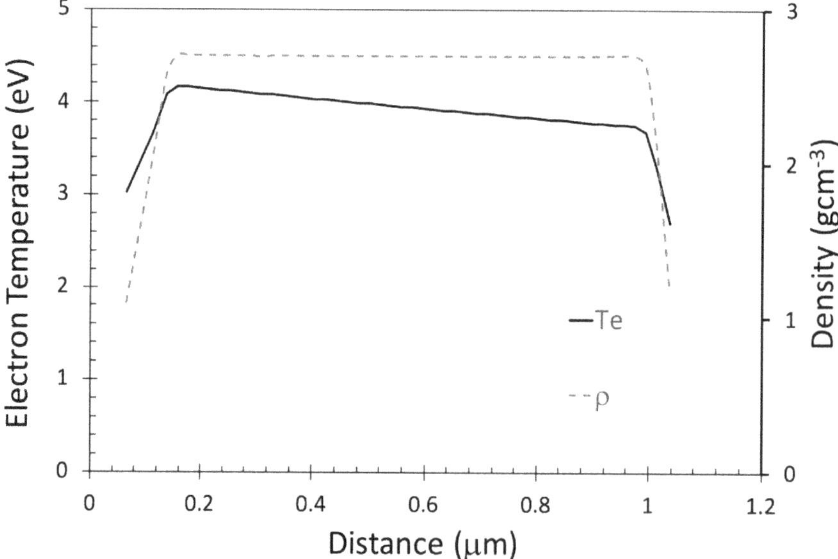

Figure 3.6. Uniform heating of an Al foil with 3.1 keV x-rays with 100 fs pulse width (FWHM). The simulation time is 5 ps after heating for an initially 1 μm thick foil.

we see just how uniform a sample can be. In this simulation, a 3.1keV x-ray pulse of 100 fs FWHM is incident on a 1 micron thick Al foil at an intensity of 10^{15} W cm^{-2}. The heating is uniform across the foil, with the maximum range of temperature being only around 10%. Of course, this advantage is offset, to a degree, by the relatively small available energy, typically in the millijoule range, meaning that sample sizes are small. In the example, of Lévy *et al*, the focal spot is less than 20 μm across and, for the simulation of figure 3.6, assuming an energy of 3 mJ, typical of a large XFEL facility, the focal spot would have to be limited to 60 μm to achieve the required intensity.

In the profile shown, we can see that, even several ps after illumination, the hydrodynamic motion has only affected the edge regions, with the bulk maintaining solid density. This is not surprising because, as discussed in chapter 1, we can expect a sound speed of order 13 km s^{-1} and so, for a 1 μm foil, we expect a disassembly timescale of order 80 ps. As we also discussed in chapter 1, Mazevet *et al* [28] have explored the timescale for structural changes in Au foils that are heated, effectively instantaneously, to energy densities appropriate to WDM and found evolution timescales of more than a few picoseconds. Clearly, the use of XFELs, where the heating pulse is of the order of an inverse phonon frequency, has a key potential benefit in allowing us to carry out experiments where we can think about probing time-evolution and equilibration effects but maintain a uniform solid density.

However, for the very fastest timescale experiments, we might need to consider the way in which energy is absorbed. This is likely to be by photo-absorption of XFEL photons by electrons in the sample. Even for absorption by the K-edge, we may end up with photo-ejected electrons at several keV initial energy. The range of such electrons in solids has been investigated, e.g. [29] and, in the 1–10 keV regime, for mid-Z elements, we expect the range to be of order 1–10 nm, with stopping times of more than a few fs. Thus, we might need to consider this in the interpretation of experiments looking at ultra-fast phenomena.

3.2 Proton and heavy ion heating

3.2.1 Ion stopping in matter

The manner in which fast ions interact with matter leads to a range of possible applications, from hadron therapy for cancer to production of doped semiconductor junctions, and, of course, the generation of WDM samples. In order to understand some important points, let us consider a simple model of a heavy particle, (which for our purposes can be a proton, α-particle, or heavier ion), with the electrons in a solid target. Consider the situation in figure 3.7.

Let us assume a projectile with a net charge of Z_ie and an atomic mass of M enters the target material with a velocity, v, which, for now, we assume to be non-relativistic. If we consider the interaction with a particular electron with an impact parameter, b, as shown in the figure, then, we know the magnitude of the force between them at closest approach, is given by;

$$F = \frac{Z_ie^2}{4\pi\epsilon_0 b^2} \tag{3.7}$$

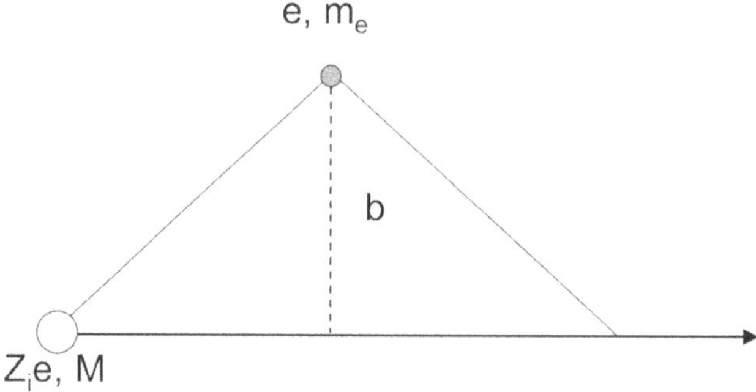

Figure 3.7. Simplistic schematic of a heavy particle interaction with the electrons in a solid target.

Since the projectile is much more massive than the electron, we assume its trajectory is more or less unaffected by the interaction and the electron recoils (in the direction of the projectile if it is positively charged) taking up some kinetic energy, which is lost from the projectile. The momentum imparted to the electron is the integral over time of the force applied. By considering that the longitudinal forces cancel out and that it is only the transverse force that is effective, we can show that the impulse is given by the force at closest approach acting for an effective time of $dt = 2b/\mathrm{v}$. The non-relativistic momentum imparted to the electron is then, Fdt and thus;

$$p_e = \frac{2Z_i e^2}{4\pi\epsilon_0 b \mathrm{v}} \tag{3.8}$$

The kinetic energy lost by the projectile due to this interaction is then;

$$K_e = \frac{2Z_i^2 e^4}{(4\pi\epsilon_0)^2 m_e b^2 \mathrm{v}^2} \tag{3.9}$$

If the target material has a density of atoms given by N_a and an atomic number of Z, then there are $2\pi b db N_a Z$ electrons within an annulus of radius b and width db. As the projectile passes through the solid it has a similar interaction with other electrons and we can integrate over all values of the impact parameter to give an instantaneous rate of loss of energy as;

$$-\frac{dK_e}{dz} = \frac{Z_i^2 e^4}{(4\pi\epsilon_0)^2 m_e} \frac{4\pi Z N_a}{\mathrm{v}^2} \int_{b_{\min}}^{b_{\max}} \frac{db}{b} = \frac{N_a Z Z_i^2 e^4}{4\pi\epsilon_0^2 m_e} \frac{1}{\mathrm{v}^2} ln\left[\frac{b_{\max}}{b_{\min}}\right] \tag{3.10}$$

As we see, there is an inverse dependence on the loss rate with the square of the projectile velocity, and thus with the energy. The upper and lower values for the effective impact parameters are usually dealt with by considering the maximum and minimum energy transfer. In order to recoil from the interaction with the ion, we can consider that the minimum energy gain, for an average atomic electron, must be the average ionisation energy, I_{avg}, for the sample material. This sets the value for $b_{\max}$, and considering equation (3.9), we can, see that $b_{\max}^2 \propto \frac{1}{I_{\mathrm{avg}}}$. For the maximum energy loss, we can set this according to a head on collision, between a light and much heavier particle, where the transfer of momentum is $\sim 2p_e$. This gives an energy loss to the ion of $\sim 2m_e v^2$ and we can see that $b_{\min}^2 \propto \frac{1}{2m_e v^2}$. Combining these two, we replace the logarithmic term in equation (3.10) with a (non-relativistic) term;

$$\frac{1}{2}\ln\left(\frac{2m_e v^2}{I_{\mathrm{avg}}}\right) \tag{3.11}$$

The treatment above is very simplistic but shows some of the key behaviour included in the Bethe stopping power equation which is given, in the relativistic version, by;

$$\frac{dK_e}{dz} = -\frac{N_a Z Z_i^2 e^4}{4\pi\epsilon_0^2 m_e c^2}\frac{1}{\beta^2}\left[ln\left(\frac{2m_e c^2\beta^2}{I_{\mathrm{avg}}(1-\beta^2)}\right) - \beta^2\right] \tag{3.12}$$

As we can see, the difference from our simple derivation is contained within the natural logarithm term. Following Bloch, this is often approximated using $I_{\mathrm{avg}} = 11Z$ eV, and using this, equation (3.12) is sometimes called the Bethe–Bloch formula. This formula is generally valid at high energies, in the MeV range, and there are several corrections that have been made, that deal with lower energy projectiles; see Ziegler [30] for a review of this subject.

The main result of our derivation, so far, is the observation that the rate of loss of energy is inversely proportional to the projectile's kinetic energy. This means that an ion initially entering a target may lose energy only slowly but then gradually loses energy more rapidly until there is a sudden, total loss of energy, at the so-called Bragg peak. We can see an example of this in figure 3.8(a), for the stopping of protons in aluminium. In this case, the stopping power data [31] has been taken from detailed calculations with the PSTAR program, tabulated on the NIST (National Institutes of Science and Technology) website.

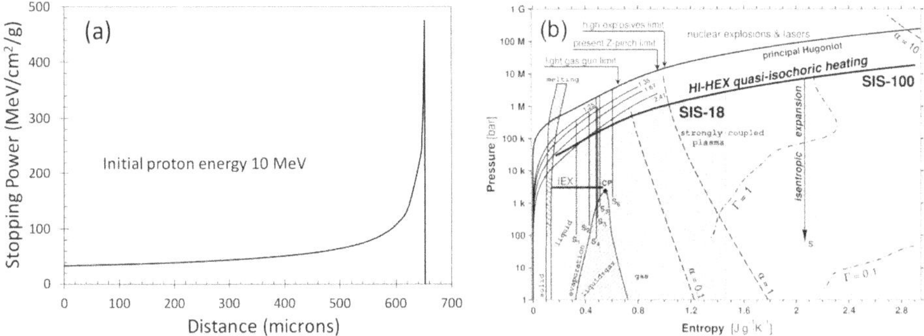

Figure 3.8. (a) We can see the stopping power of protons with 10 MeV initial kinetic energy, as a function of depth into an Al sample. This graph was produced using the stopping power data from the NIST database [31]. (b) Illustration of the potential for WDM creation of the upgraded SIS-100 ion beam facility at GSI Darmstadt. Figure reprinted with permission from Tahir *et al* [32]. © The Royal Swedish Academy of Sciences. All rights reserved. We can see that, for Pb in this case, we can reach the multi-Mbar WDM regime.

As we discussed in chapter 2, ion beams at hundreds of MeV per atomic mass unit can be generated in heavy ion beam facilities and the range of these ions can be several mm even in high *Z* samples. This provides a significant advantage over x-ray heating, in that mm scale higher *Z* samples of WDM can be created, where x-rays in the keV range would struggle to penetrate and, at least for laser-plasmas, harder x-ray sources would rapidly decrease in efficiency. There are several proposals for exploiting this possibility. One example is the HIHEX proposal [32, 33] whereby an ion beam with a diameter of up to 2–3 mm can be used to isochorically heat a ~mm diameter cylindrical plasma, allowing deposition of 100s MJ/kg along a several mm long sample. The potential for WDM creation is summarised in figure 3.8(b) for the case of uranium ions and a lead target [32]. This potential is not limited to high *Z* targets, for example, Tauschwitz *et al* [34] have discussed the isochoric heating of solid density hydrogen to several eV in a mm scale target using the same facility.

As we can see, a key advantage of ion beam WDM generation is that we can create large samples with high *Z* materials, that would not be easily penetrated by an x-ray source. However, this also means that we cannot probe the sample with x-rays, and some consideration would have to be given to the diagnostics used to characterise the samples created.

3.2.2 Laser driven proton beam heating

Since the early 2000s, it has been known that irradiation of a solid foil target at intensities of 10^{18} W cm^{-2} and above can lead to the generation of intense beams of protons with energies of order 10s of MeV [35, 36]. The principal mechanism by which this occurs is Target Normal Sheath Acceleration (TNSA). In this mechanism, a beam of fast electrons is created on the laser irradiated side of the foil. As the electrons pass through the rear side of the foil they pick up protons that come from hydro-carbon contaminants on the surface of the foil. A significant experimental feature is that the protons are generated in a spectrum. An example can be seen in figure 3.9(a) in which we can see that multi-MeV protons can be created even with a

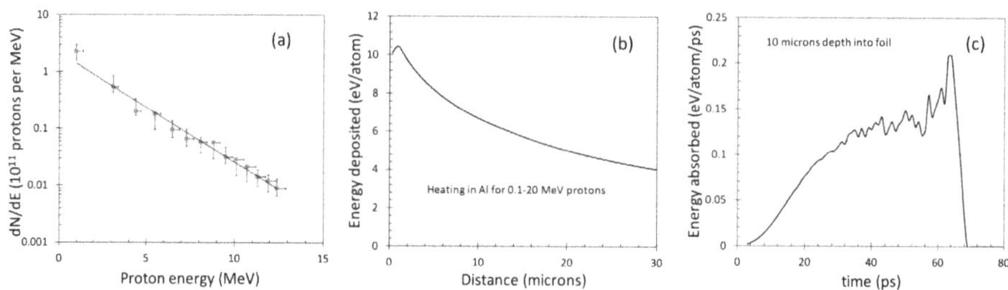

Figure 3.9. (a) Spectrum of protons from an experiment carried out with the TARANIS laser at Queen's University Belfast. The laser energy on target was 7.5 J at 1.053 μm and the focussed intensity is around 10^{18} W cm^{-2} onto a 6 μm thick Al foil. (b) Simulated heating profile for an Al foil target. We assume a spectrum similar to the one given in figure 3.9(a), but have extrapolated the distribution to protons of 0.1–20 MeV energy and assumed a larger laser facility with 50 J laser energy. (c) Time history of heating for 10 μm depth into the Al foil, for the same simulation as in (b).

relatively small laser system. With a larger system, spectra similar to this but extending to even higher proton energies can be seen.

By now, this is a routine experimental technique that has been well studied. Conversion efficiencies from laser light of up to 10%, with proton energies in excess of 85 MeV, have been achieved [37]. In addition to their use as a diagnostic probe and proposed applications as the basis of a hadron therapy facility, such proton beams have also been proposed as a way of heating a secondary target to WDM conditions. The energy deposition of 10 MeV protons that we saw in figure 3.8(a) shows us that we can have a relatively uniform deposition over a significant depth of target material as long as we work well within the Bragg peak. However, for heating of WDM samples, we should take into account that there is a spectrum of protons, as seen in figure 3.9(a), and we do this in the simulations shown in figure 3.9 and discussed below.

The proton beams generated in such experiments, typically, have a divergence that depends on the proton energy, but is generally of the order of about half a radian. This means that, even at 1 mm from the source, the energy is deposited over an area 0.5 mm in diameter, thus restricting the heating level achievable. Furthermore, the closeness of the proton source and the secondary target makes it difficult to find a geometry where the sample can be probed. The, already noted, energy spread of the protons leads to another difficulty in generating WDM samples, which is the temporal spread of the protons reaching the target. If we imagine our secondary target to be as close as 1 mm from the proton source foil, then, for 2 MeV protons, the flight time is of order 50 ps, for 4 MeV protons it is ~35 ps. This spread in arrival time means that heating is not instantaneous and placing of the secondary target has to be close in order to prevent significant hydrodynamic expansion during the experiment.

We can, nevertheless, consider the sample heating that might be created with a large enough available laser energy. If we take a spectrum of protons similar to that displayed in figure 3.9(a), but assume a 50 J laser energy and a divergence of 0.5 radians, with a 1 mm offset distance, we can divide the protons into energy groups and calculate their energy loss and their energy as a function of distance (in terms of areal density), through a sample. We show the results of this simple approach in figure 3.9(b). As we can see, more heating is seen near the front, because the slower protons do not penetrate far into the foil. Despite this, we can see that for a 10 μm thick Al foil, we would achieve heating that is uniform to within about 20% of an average of about 8 eV per atom. This would lead to a pressure of over 80 GPa and a hydrodynamic expansion time of order ~1 ns. We can compare this with the time heating history for a point at 10 μm depth into the foil, seen in figure 3.9(c), where the zero of the timescale is the arrival of the fastest proton group. We can see that the effect of the spectral shape, with proton number decreasing with energy, is to generate a linearly rising heating profile. In this case, it is cut off at around 70 ps, which corresponds to the relative time of arrival for ~1 MeV protons, which are the lowest energy protons to reach this deep into the foil. For shallower depths, the heating profile will be more extended as slower proton groups take longer to arrive, but can still penetrate the foil. For many experiments, probing can take place before the slowest protons reach the sample. If this is still an issue, then it can be dealt with, for example, by tamping the sample with a different material, such as CH, thus allowing a thinner foil of sample material to be uniformly heated in a timescale under 100 ps but with a hydrodynamic expansion timescale that can be longer than this. If the tamping interferes with the diagnostics, then a moderator layer between the proton source foil and the sample can help to remove the large number of slower protons generated in the spectrum.

Because of these issues of divergence and temporal spread, experimental schemes to focus the protons have been investigated. In one of these, Patel *et al* [38] used hemispherical targets. The protons are accelerated normal to the target surface and thus the curved target created a focussing effect. Using this method, heating of the sample to about 20 eV was demonstrated. However, the sub-mm proximity of the intense laser plasma interaction at $I > 10^{18}$ W cm^{-2} complicates probing of the sample and it is only from the optical emission from the sample surface that the temperature is diagnosed. An alternative scheme, using transient electrostatic focussing via helical targets has been under development [39]. In this scheme, a foil target is attached to a helically coiled wire. As the laser interacts with the foil, the fast electrons generated can escape the target and thus a fast-rising current passes though the target, including the helical coil. A transient charge in the wire then creates an electric field that acts to focus the protons.

3.3 Fast electron heating

The TNSA mechanism described above relies on the generation of supra-thermal electrons. This can happen by different mechanisms, one of which is resonance absorption, in which a p-polarised laser pulse is incident at an angle θ from the normal to a target surface. The laser is refracted in the surface plasma and reaches a turning point density of $n_c \cos^2\theta$, where n_c is the critical density discussed in chapter 2, (see for example [40]). At this point, the electric field oscillates parallel to the density gradient, resonantly driving a plasma wave at the critical density surface. As the plasma wave damps, electrons are accelerated into the target in the direction normal to the target surface. This mechanism can occur at irradiances up to around 10^{18} W cm^{-2}. At higher irradiances, we see the acceleration change to the $J \times B$ mechanism [41] in which the electrons are accelerated in the direction of the laser.

The efficiency of conversion into such fast electrons can be 10s of percent. The effective fast electron temperature has been shown to vary as $I^{1/3}$ by several authors. One commonly used scaling is that of Beg *et al* [42];

$$T_{\text{hot}}(\text{keV}) = 100(I_{17}\lambda^2)^{1/3} \tag{3.13}$$

where $I_{17}\lambda^2$ is in units of 10^{17} W cm$^{-2}\mu$m^2. Other authors give a similar scaling but with a lower constant and sometimes a dependence on the temperature of the sub-critical coronal plasma, in which the laser propagates on the way to the turning point density. For example, Wilks and Kruer [43] give;

$$T_{\text{hot}}(\text{keV}) = 10(T_{bg}I_{15}\lambda^2)^{1/3} \tag{3.14}$$

where T_{bg} is the coronal plasma temperature and $I_{15}\lambda^2$ is in units of 10^{15} W cm$^{-2}\mu$m^2. The collisional range of such electrons in solids can be 100s of microns. However, this does not necessarily give us the depth to which fast electron heating can penetrate the solid. As discussed by Bell *et al* [44] the current generated by fast electrons alone would exceed the Alfvén limit, given by;

$$I_A(\text{kA}) = 17\beta\gamma \tag{3.15}$$

in which β and γ have their usual meaning in relativistic mechanics. For an electron beam of ~100 keV energy, this leads to a current of around 20 kA. In fact, the current of fast electrons is expected to be much higher. For example, with 100 J on target in a 1 ps pulse, if we assume a modest 20% absorption and $T_e = 100\,\text{keV}$, we expect a current of more than 100 MA. What occurs is that a nearly equal, balancing, return current of electrons at lower energy is present to provide charge neutrality and allow the fast electrons to penetrate. The return current electrons are more collisional and it is these that are responsible for resistive heating of the sample. The depth of the resistive heating will, for course, depend on how far the fast electrons penetrate, which in turn depends on their collisionality and the retarding electric field set up that drives the return current. Bell *et al* [44] have calculated that this penetration depth can be given by;

$$Z_{\text{fast}}(\mu\text{m}) = 12\left(\frac{T_{\text{hot}}}{200\text{keV}}\right)^2\left(\frac{\sigma}{10^6\Omega^{-1}\text{m}^{-1}}\right)\left(\frac{I_{\text{abs}}}{10^{18}\text{W cm}^{-2}}\right)^{-1} \tag{3.16}$$

The electrical conductivity, σ depends on the bulk temperature, as well as the material, and thus would vary during the heating process. However, if we take a typical value of 10^6 Ω^{-1} m^{-1} and absorption of 10^{18} W cm^{-2}, we see that with a fast electron temperature of 200 keV, from the Beg *et al* scaling, we would have a penetration depth of around 10 μm. This process has been used, for example by

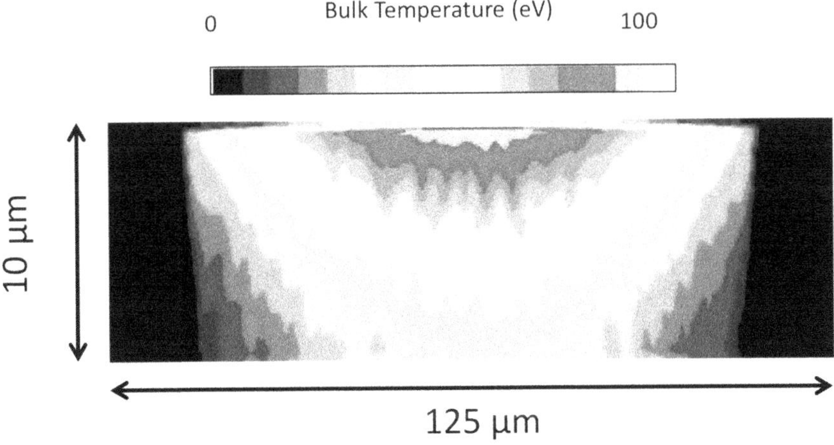

Figure 3.10. Zephyros simulation of the bulk temperature in a solid Ti foil at the end of a 1 ps pulse focussed to 2×10^{18} W cm^{-2} at 1.053 μm wavelength.

Hansen *et al* [45], to create WDM at temperatures of 10s eV and solid density, leading to ion–ion coupling parameters of $\Gamma > 1$ and partial degeneracy. A simulation of such an experiment is shown in figure 3.10. In this case, we have used a 1.053 μm wavelength, 1 ps duration pulse incident on a 10 μm thick Ti foil at $\sim 2 \times 10^{18}$ W cm^{-2}. We expect the effective fast electron temperature to be ~250 keV. The simulation has been carried out with the Zephyros code [46], which is a 3-D hybrid code that models the transport of fast electrons injected into a solid by a high power laser-plasma interaction. The code uses the Lee-More model of the conductivity [47] to calculate the expected return current and resistive heating and calculates the E and **B** fields generated by the current of fast electrons. As we can see, the resistive heating, in this case, leads to a solid density sample at temperatures 10s of eV. In the next chapter we will see how, in this kind of sample, the fast electrons generate K-α emission that acts as a diagnostic of the temperature.

Even more so than the proton heating discussed above, a potential limitation of this technique is that the WDM sample is necessarily in close proximity to a very high temperature plasma created by the intense laser interaction and this can create problems with hard x-ray background in diagnostics as well as electromagnetic pulse interference with electronics, which is a known problem with intense laser-plasma experiments. In addition to this problem, we also have strong gradients in the temperature that would complicate the task of diagnosing the WDM conditions.

The possibility of using energetic electron beams, from an accelerator, for high energy density science, avoiding the use of a laser-plasma, has been discussed theoretically [48]. Kikuchi *et al* [49] propose an optimum design, for WDM experiments, that uses a 100 ns duration, 1 MeV electron beam, with ~10 kA current, to heat a mm scale Al block to ~1 eV. The long pulse duration indicates that careful design would be necessary to avoid hydrodynamic expansion during the experiment. Experiments with this type of facility are, at present, relatively uncommon. However, for example, Coleman *et al* [50] have used 19.8 MeV electrons, in a 100 ns bunch, to isochorically heat 0.2 mm thick Cu foils to WDM conditions of up to around 1 eV. Hydrodynamic disassembly was seen to occur about half-way through the heating where the pressure was about 20 GPa.

An important aspect of WDM creation with fast electrons is that, even though experiments might not be optimal for studying WDM itself, the generation of this state is an important part of understanding the process of fast electron ignition [51] which is still a matter of considerable interest in the laser-plasma community.

References

[1] Batanov V A, Gochelashvili K S, Ershov B V, Malkov A N, Kolisnichenko P I, Prokhorov A M and Fedorov V B 1974 *JETP Lett.* **20** 185–7

[2] Burkhalter P G and Nagel D J 1975 *Phys. Rev.* A **11** 782–8

[3] Boiko V A, Faenov A Y and Pikuz S A 1978 *J. Quant. Spectrosc. Radiat. Transf.* **19** 11–50

[4] Sigel R, Eidmann K, Lavarenne F and Schmalz R 1999 *Phys. Fluids* B **2** 199

[5] Kania D R *et al* 1992 *Phys. Rev.* A **46** 7853

[6] Goldstone P 1987 *Phys. Rev. Lett.* **59** 56–9

[7] Kornblum H N, Kauffman R L and Smith J A 1986 *Rev. Sci. Instrum.* **57** 2179–81

[8] Matthews D L 1982 *J. Appl. Phys.* **54** 4260–8
[9] Griem H R 1964 *Plasma Spectroscopy* (New York: McGraw-Hill)
[10] Salzmann D 1998 *Atomic Physics in Hot Plasmas* (Oxford: Oxford University Press)
[11] Karzas W J and Latter R 1961 *Astrophys. J. Suppl.* **6** 167
[12] Garban-Labaune C, Fabre E, Max C E, Fabbro R, Amiranoff F, Virmont J, Weinfeld M and Michard A 1982 *Phys. Rev. Lett.* **48** 1018–21
[13] Nishimura H *et al* 1991 *Phys. Rev.* A **44** 8323–833
[14] Farmer W A, Tabak M, Hammer J H, Amendt P A and Hinkel D E 2019 *Phys. Plasmas* **26** 032701
[15] Henke B L, Gullikson E M and Davis J C 1993 *At. Data Nucl. Data tables* **54** 181–342
[16] Rybicki G B 1979 *Radiative Processes in Astrophysics* (New York: Wiley)
[17] Apruzese J P, Davis J, Whitney K G, Thornhill J W, Kepple P C, Clark R W, Deeney C, Coverdale C A and Sanford T W L 2002 *Phys. Plasmas* **9** 2411–9
[18] Dewald E L *et al* 2008 *Phys. Plasmas* **15** 072706
[19] Phillion D W and Hailey C J 1986 *Phys. Rev.* A **34** 4886–96
[20] Kettle B *et al* 2015 *J. Phys. B: At. Mol. Opt. Phys.* **48** 224002
[21] Hu G-Y *et al* 2008 *Laser Part. Beams* **26** 661–70
[22] Back C A *et al* 2001 *Phys. Rev. Lett.* **87** 275003
[23] Glenzer S H, Gregori G, Lee R W, Rogers F J, Pollaine S W and Landen O L 2003 *Phys. Rev. Lett.* **90** 175002
[24] Kettle B *et al* 2016 *Phys. Rev.* E **94** 023203
[25] Lee R W *et al* 2003 *J. Opt. Soc. Am.* B **20** 770–8
[26] Zastrau U *et al* 2008 *Phys. Rev.* E **78** 066406
[27] Levy A *et al* 2015 *Phys. Plasmas* **22** 030703
[28] Mazevet S, Clérouin J, Recoules V, Anglade P M and Zerah G 2005 *Phys. Rev. Lett.* **95** 085002
[29] Ashley J C, Tung C J, Ritchie R H and Anderson V E 1976 *IEEE Trans. Nucl. Sci.* **NS-23** 1833–7
[30] Ziegler J F 1999 *J. Appl. Phys.* **85** 1249
[31] Andersen H H and Ziegler J F 1977 *The Stopping and Ranges of Ions in Matter* vol 3 (Elmsford, NY: Pergamon)
[32] Tahir N A, Spiller P, Piriz A R, Shutov A, Lomonosov I V, Schollmeier M, Pelka A, Hoffmann D H H and Deutsch C 2008 *Phys. Scr.* **T132** 014023
[33] Mintsev V *et al* 2016 *Contrib. Plasma Phys.* **56** 281–5
[34] Tauschwitz A, Maruhn J A, Riley D, Shabbir N G, Rosmej F B, Borneis S and Tauschwitz A 2007 *High Energy Density Phys.* **3** 371–8
[35] Snavely R A *et al* 2000 *Phys. Rev. Lett.* **85** 2945–8
[36] Clark E L *et al* 2000 *Phys. Rev. Lett.* **85** 1654–7
[37] Wagner F *et al* 2016 *Phys. Rev. Lett.* **116** 205002
[38] Patel P K *et al* 2003 *Phys. Rev. Lett.* **91** 125004
[39] Ahmed H *et al* 2016 *Sci. Rep.* **7** 10891
[40] Kruer W L 2003 *The Physics of Laser Plasma Interactions* (Boulder, CO: Westview Press)
[41] Wilks S C, Kruer W L, Tabak M and Langdon A B 1997 *Phys. Rev. Lett.* **69** 1383–6
[42] Beg F N *et al* 1997 *Phys. Plasmas* **4** 447
[43] Wilks S C and Kruer W L 1997 *IEEE J. Quantum Electron.* **33** 1954–68
[44] Bell A R, Davies J R, Guerin S and Ruhl H 1997 *Plasma Phys. Control. Fusion* **39** 653–9

[45] Hansen S B *et al* 2005 *Phys. Rev.* E **72** 036408
[46] Robinson A P L 2014 Zephyros User Manual *Technical Report* RAL-TR-2014-013
[47] Lee Y T and More R M 1984 *Phys. Fluids* **27** 1273–86
[48] Joshi C *et al* 2002 *Phys. Plasmas* **9** 1845–55
[49] Kikuchi T, Sasaki T, Horioka K and Harada N 2009 *J. Plasma Fusion Res: Rapid Commun.* **4** 026
[50] Coleman J E, Morris H E, Jakulewicz M S, Andrews H L and Briggs M E 2018 *Phys. Rev.* E **98** 043201
[51] Tabak M, Hammer J, Glinsky M E, Kruer W L, Wilks S C, Woodworth J, Campbell E M, Perry M D and Mason R J 1994 *Phys. Plasmas* **1** 1626
[52] Robey H F, Perry T S, Park H-S, Amendt P, Sorce C M, Compton S M, Campbell K M and Knauer J P 2005 *Phys. Plasmas* **12** 072701

IOP Publishing

Warm Dense Matter
Laboratory generation and diagnosis
David Riley

Chapter 4

X-ray diagnostics

As we have seen in the previous two chapters, x-ray drives are important in the generation of warm dense matter (WDM) samples, both by shock generation and by volumetric heating. Thus, it is relevant here to discuss some of the techniques and equipment needed to characterise x-ray sources. Equally importantly, whilst optical techniques can only generally probe the surface of a warm dense sample, x-rays have the ability to penetrate a dense, relatively cool sample and so form the basis of diagnostic techniques such as x-ray scattering and absorption spectroscopy. The combination of high density and relatively modest temperature found in WDM means that x-ray emission spectroscopy is often of limited value. However, we will also note some exceptional cases, where non-thermally generated x-ray emission may be used to provide information on the temperature and ionisation balance within a sample.

4.1 X-ray dispersion and detection

In this section, we will discuss some experimental considerations involved in the spectral dispersion and detection of x-rays. This is essential to more or less all of the diagnostic techniques we shall discuss.

4.1.1 Bragg crystal spectrometers

For x-rays in the 1–10 keV photon energy range, the most suitable method for spectrally dispersing emitted or scattered x-rays is use of a Bragg crystal spectrometer. We can see a schematic of Bragg diffraction in figure 4.1. The basic equation of interest [1] is;

$$n\lambda = 2d \sin \theta_B \tag{4.1}$$

where the integer, n is the order of reflection, λ is the wavelength, d is the interplane distance, which depends on the crystal and the crystallographic cut and $2\theta_B$ is the angle through which the x-rays are deflected in the diffraction process.

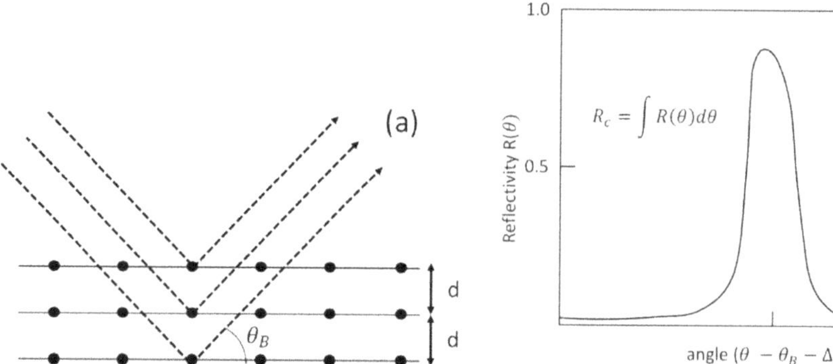

Figure 4.1. (a) Schematic of the Bragg diffraction of x-rays from a crystal with inter-planar distance, d. Constructive interference is achieved if the path length between x-rays scattered from atoms in adjacent planes is an integer number of wavelengths, leading to the well-known Bragg condition (equation (4.1)) (b) Schematic of a rocking curve, showing how the reflectivity varies with angle. The peak is slightly shifted from the Bragg angle by the small term, Δ, which accounts for the refractive index of the material in the x-ray regime.

In practice, the reflectivity of x-rays, of a particular wavelength, occurs over a small range of angles, to give a *rocking curve*, usually of order 10^{-4} radians in width, as shown in figure 4.1(b). The width of the rocking curve determines the ultimate resolution, $E/\Delta E$, that will be possible in a crystal spectrometer. This can be of order $10^4 - 10^5$, though in practice, detector resolution and source broadening will prevent such high resolution, in most cases. We can estimate a typical source and detector broadening as follows. Looking at figure 4.1, we can imagine that the x-rays, for a particular Bragg angle, travel a total distance, L, from the source, via the crystal, onto a detector surface to the right. If the detector is normal to the incoming rays, then a change in angle, $d\theta$, leads to a change is position on the detector, $ds = Ld\theta$ and the dispersion, for $n = 1$, is given, from equation (4.1), by;

$$\frac{d\lambda}{ds} = \frac{2d\cos\theta_B}{L} \tag{4.2}$$

We can use this equation, by equating ds with a typical pixel or resolution element on our detector, to determine the wavelength resolution as $d\lambda$. Alternatively, we can consider a symmetric argument and consider ds to be the projected size of the source, normal to the x-rays emitted towards the crystal, to calculate the source broadening with the same equation. For a laser plasma backlighting experiment, we might have a source size of typically 100 μm and a source-detector distance of ~30 cm. If we assume a typical $2d = 10$ Å and a Bragg angle of 30°, we can calculate a resolution, due to source broadening, of ~1700. Usually the resolution of the detector can be better than this, but should also be folded in.

The integrated reflectivity, R_c, determines the signal level and, as with reflection of optical wavelengths from a surface, the reflectivity depends on the polarisation. For p-polarised x-rays, reflection is minimised at the Brewster angle, given by $\arctan(n_r)$ where n_r is the refractive index. Since $n_r \sim 1$ for the x-ray region, the Brewster angle is close to 45° and the p-polarised x-ray reflection is close to zero.

It can be important, in some experiments, to make sure this is accounted for, especially if scattering of polarised x-rays, for example from an x-ray free electron laser, is being considered.

For some crystal types, the structure is not perfect but consists of small crystallites that have a spread of orientations about an average. This is called a mosaic crystal and they display a broadening of the rocking curve and a generally higher integrated reflectivity. A particularly significant example is highly-oriented pyrolytic graphite (HOPG) [2–4]. Depending on the grade, the spread in crystallite orientations can be from approximately 0.1–1° with integrated reflectivity of order a few times 10^{-3} radians. An important restriction with the use of HOPG is that, due to the phenomenon of mosaic focussing [3] (sometimes called parafocusing), in order to have good spectral resolution, the source to crystal and crystal to detector distances should be comparable, ideally the same. With this condition satisfied, resolutions of several thousand can be obtained [4]. An additional feature of HOPG is that it does not need to be grown in bulk, as many crystals do, but can be deposited onto a substrate to form a layer that may be typically of order 100 μm thick. The substrate need not be flat, but could be a curved surface, to create a focusing crystal. In table 4.1, we list some properties of commonly used crystals and the Miller indices of the cut used. We give the typical reflectivity for both the perfect and mosaic crystals at an assumed Bragg angle of 30°. Fuller details including reflectivity across all wavelengths can be found, for example, in [5].

Focussing crystals

Thus far, we have had in mind flat crystals, as shown in figure 4.1. However, in order to gather more signal and/or to provide spatial resolution, we can use curved crystals. We start by considering cylindrical focussing in the von-Hamos configuration. As we can see in figure 4.2(a), the von-Hamos configuration concentrates the diffracted signal into a line-focus that provides an enhanced signal to noise ratio. In principle, this arrangement also provides 1-dimensional spatial resolution of the emitting source. In practice, how practical this is will depend on the source

Table 4.1. Some parameters of crystals commonly used for x-ray spectroscopy in warm dense matter experiments. The perfect and mosaic crystal reflectivities, R_p and R_m are evaluated for a typical Bragg angle of 30°. The typical spectral ranges assume a practical range of Bragg angles of 15–45°. For longer wavelengths, where the photon energy drops below about 1 keV, strong absorption by filters may become an experimental limitation.

Crystal	$2d$ (Å)	R_p (mrad)	R_m(mrad)	Range (Å)
Silicon (220)	3.84	0.077	0.185	2.7–3.7
Silicon (111)	6.271	0.053	0.114	4.4–6.0
Graphite (0002)	6.708	0.11	2.5	4.7–6.5
PET (002)	8.742	0.08	0.55	6.2–8.4
ADP (101)	10.64	0.13	0.16	7.5–10.3
KAP (001)	26.634	0.023	0.03	19–25

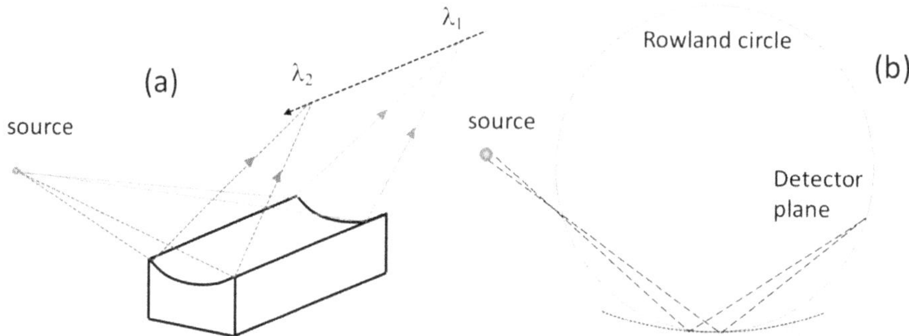

Figure 4.2. (a) Schematic of a von-Hamos crystal arrangement. If the source is placed at a height above the crystal centre equal to it's radius of curvature then spatial resolution in the direction perpendicular to the wavelength dispersion can be achieved. (b) Johann geometry. In this configuration, source broadening can be substantially eliminated as long as the detector is on the Rowland circle and resolutions of ∼5000 have been achieved [6].

size, resolution of the detector and the optical quality of the crystal surface. The highly mosaic nature of HOPG generally makes it less suitable for high resolution spatial imaging. However, its ability to provide high integrated reflectivity with good spectral resolution and to be readily formed into a focussing geometry makes it especially useful in applications such as x-ray Thomson scattering, where a low signal may need to be distinguished from a significant level of background emission.

A cylindrically curved crystal can also be used in the Johann configuration, see figure 4.2(b). In this configuration, we note that the crystal with radius of curvature, R, just touches the edge of the Rowland circle, which has a radius $2R$. An x-ray emitted from the source will pass through a point on the Rowland circle on its way to the crystal. The Bragg angle formed at the crystal results in the x-rays being projected through a symmetric point on the Rowland circle. As we see in the diagram, the curvature of the crystal means that a ray of the same wavelength, emitted from a different part of the source, will meet the necessary Bragg angle at a different part of the crystal but will pass through the same point on the Rowland circle and be projected through the same symmetric point. This means that source broadening can be effectively eliminated, allowing high spectral resolution, even for relatively large source sizes, which can be especially important if we are dealing, for example, with mm sized samples probed by x-ray scattering.

In fact, the Johann configuration does not achieve perfect spectral focussing. To achieve this, the Johansson configuration is used. In this, the curvature of the crystal is still twice the Rowland circle radius but now the surface of the crystal is machined to lie on the Rowland circle along its entire length. This is a much less used configuration in plasma spectroscopy as the additional machining is non-trivial.

For some applications, we can obtain high spatial and spectral resolution with spherically focussing crystals [7], which are available for a variety of crystal cuts.

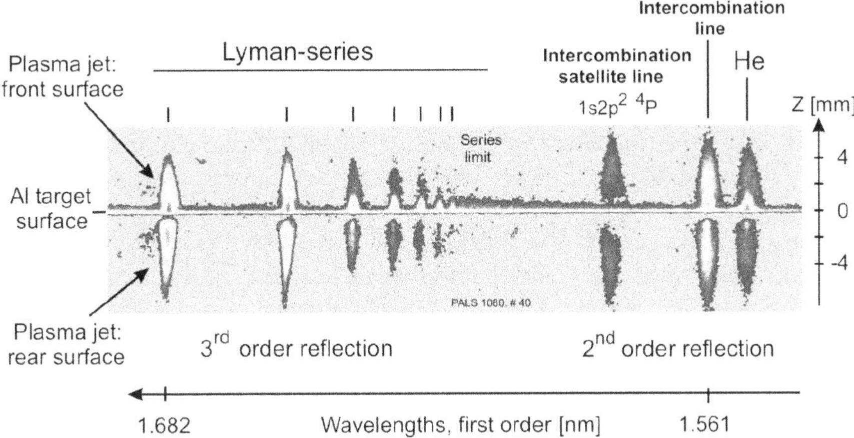

Figure 4.3. X-ray spectrum of He-like and H-like emission from an Al plasma using a spherical mica crystal. The reflections are in 2nd and 3rd order. Although the exploding foil is not a WDM sample, the image nicely illustrates the combination of spatial and spectral resolution possible with a spherical crystal. Despite the large plasma size, resolution of order E/ΔE $\sim 10^3$ has been achieved. Image reproduced with permission from [8] ©2006 EDP Sciences.

Alternatively, we can make a quasi-monochromatic 2-D image. In the latter case, the best focus is given by the equation;

$$\frac{1}{u} + \frac{1}{v} = \frac{2 \sin \theta_B}{R} \tag{4.3}$$

where u is the distance from the source to the centre of the crystal surface, v is the distance from the crystal to the image plane and R is the crystal radius of curvature. For simultaneous spectral resolution and 1-D spatial resolution in the sagittal direction, we can place the detector onto the Rowland circle, where the Johann focusing removes the contribution of source size to spectral broadening. In this case, the geometry required sets the distance to the source, u as;

$$u = \frac{R \sin \theta_B}{(2 \sin \theta_B - 1)} \tag{4.4}$$

In figure 4.3 we can see an example spectrum taken using a mica crystal, which has $R = 150$ mm and $2d = 19.8$ Å. The spectrum shows emission from an exploding Al foil irradiated with a 438 nm wavelength, ~250 ps duration, laser pulse with intensity $\sim 10^{15}$ W cm^{-2}, which creates a rapidly expanding hot foil, emitting He-like and H-like lines.

4.1.2 Electronic x-ray detectors and image plates

A widely used detector for keV x-rays is the CCD (charge-coupled device). These have been reviewed in the literature extensively, for example by Holst and Lomheim [9], but we give a brief overview here.

A CCD works by utilising the properties of a silicon pn junction. If a positive potential is applied to the n-type Si, the electron (negative) carriers are pulled away from the junction region, and the positive charge carriers (holes) in the p-type Si are repelled. This creates a potential well region (depletion layer), at the junction, that is free of carriers. A photon absorbed in this region will liberate electrons which will be trapped in the potential well. In the keV photon energy regime, the ejected electrons will have substantial energy and will collide with other electrons, promoting them to the conduction band as well. The band gap for Si is 1.1 eV but energy is also lost, for example, to the lattice via collisions and in this regime approximately 3.65 eV is required per electron liberated. Thus, a 1 keV photon will create roughly 300 electrons. With 16-bit analogue to digital (A-D) conversion, sensitivity of 2–3 electrons per count is possible and the CCD can be sensitive to single photons. With a typical well capacity in the regime of 10^5 electrons, a high dynamic range is possible.

Once the charge has been collected in the wells, it is moved to the readout circuitry by transferring the charge from one potential well to the next. This is achieved in different ways depending on the model, with two, three and four phase clocking all possible. In this, each pixel area has 2–4 polysilicon electrodes, which are switched high and low in a sequence that forces the charge in the direction of the read-out register and from one pixel to another, without mixing the charges stored in each pixel. The fidelity with which the charge is preserved depends on the clock speed and is generally extremely high (>99.9 %). Clock speeds in scientific grade CCDs are often slower than in video capture CCDs but speeds of ~5 MHz are available. For a CCD of 1024 × 1024 pixels, this means the read-out time is ~1 s. For many WDM experiments, the shot rate is much slower than this, limited by the repetition rate of the heating source or by the need to replace target samples.

Typical pixel sizes are of order 10–30 μm, for CCDs used in x-ray detection. However, this does not guarantee this level of resolution, as the charge cloud, created by absorption of an x-ray photon outside the potential well depth, is not well constrained and can spill over into adjoining pixels, to create *split events*. A typical CCD chip has a depletion layer depth of about 7 μm (for instance, the EEV-1511 chip). We can get a rough estimate of the likelihood of split events by calculating the fraction of photons of a given energy that will be absorbed in Si over this distance [10]. For example, a photon of 3 keV energy has an absorption length of 4.4 μm and so we expect 80% of photons to be absorbed in the depletion region and the rest contribute to *split events*.

In experiments, the ability to count photons individually has been used when the signal level is expected to be low and good spectral resolution is not needed. For example, figure 4.4(a) shows data from an early experiment to measure angularly resolved x-ray scattering from an Al WDM sample. The scattered x-rays are detected directly by the CCD. We can see, in the figure, a typical histogram of the CCD. There are several features of note. The strong peak centred around zero counts is due to the thermal noise. This arises because a background is taken before the data shot and, since the thermal noise is random, pixels in the data shot may have more or fewer thermal electrons than the background. Thus, when background subtraction occurs, for pixels with no signal photon, we see a Gaussian distribution

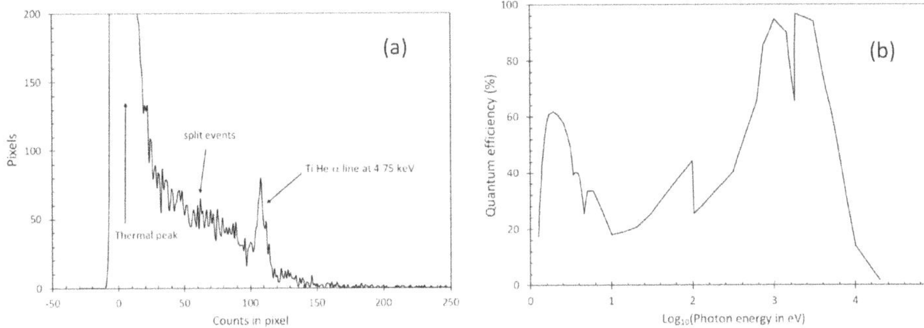

Figure 4.4. (a) Example of single photon counting histogram (see [11] for similar data). In this case, the source is scattered Ti He-α radiation at 4.75keV from a WDM sample. (b) Quantum efficiency (QE) for a back-thinned CCD.

of full-width-half maximum, typically equivalent to around 200 eV in photon energy. We can also see the peak created when pixels absorb the scattered He-α photons at 4.75 keV. The distribution of signal between this peak and the thermal peak arises partly from plasma background emission but largely from split events.

As we can see in figure 4.4(b), a CCD camera generally drops off in sensitivity above about 8 keV and even use of a deep depletion chip, where the depletion layer can be a few 10s μm thick, does not generally extend utility much beyond about 10 keV. For higher photon energies, it may be desirable to move to indirect detection, using a scintillator to convert x-rays to optical photons which are delivered to the detector via fibre-optic coupling. In experiments with samples irradiated by intense high power lasers, unless careful steps are taken to provide screening, the CCD can be susceptible to electromagnetic pulse noise [12]. Mitigating steps can be taken, for example housing the CCD in a Faraday cage, constructed from a grounded metallic enclosure.

Other electronic detectors

CCD technology has dominated scientific imaging for some time, but, in recent years, CMOS (Complementary Metal Oxide Semiconductor), technology, which was developed around the same time, has seen a resurgence, as a result of improved manufacturing capabilities. The physics is essentially similar, with comparable pixel sizes, but each pixel has its own independent digitisation and amplification. This makes it, in principle, less uniform in response, but massively parallel and capable of read-out speeds hundreds of times faster than an equivalent sized CCD array. Harada *et al* [13] have tested such a device, for direct x-ray detection, and found quantum efficiency of over 90% for up to 1 keV photons. Another key advantage is that whilst, in a CCD, the charges pass through lines of pixels to be read out, image smearing could potentially occur, depending on the duration of the x-ray source. With CMOS technology, the independence of each pixel means that a global shutter can be electronically applied, to capture an image without smearing. The presence of so much on-chip processing circuitry means that one issue that may affect CMOS more than CCD technology in direct x-ray detection is durability due to high energy photon damage.

Over the last decade, very large area detectors, based on CMOS technology, such as the Cornell-SLAC hybrid Pixel Array Detector (CSPAD) [14, 15] have been developed for use at x-ray free electron facilities [16], where the fast read-out rate available with CMOS technology is very desirable. These are based on arrays of silicon diodes, typically 194 × 185 pixels of size 110 μm that can be tiled to create detectors with areas in excess of 10 cm × 10 cm. The thick silicon layer means high quantum efficiency is maintained to higher photon energies [14].

Image plates

An alternative type of detector, that has found wide use, is the image plate [17]. In these, the principal feature is a phosphor layer typically up to a few hundred microns thick. This is usually BaF(X):Eu $^{2+}$ where X = Br, I with the ratio of Br to I depending on the model of image plate. The x-ray photon ionises the Eu $^{2+}$ to Eu $^{3+}$ and the photo-ejected electron finds its way into an F-centre trap level in the crystals of the phosphor. The trap level is metastable and the electron will generally stay there unless it is thermally excited back into the conduction band, from where it can then recombine with Eu $^{3+}$. This process is called fading and needs to be accounted for in processing the image plate to extract the data [18]. The latter is carried out in the image plate reader by focusing a He–Ne laser onto the surface in a tight focal spot. This photo-stimulates the trapped electrons which can then recombine, emitting characteristic optical photons (~390 nm), which are then detected by a photo-diode. The signal collected is proportional to the number of liberated electrons and thus to the number of incident x-rays. A key feature is that they have a peak response per photon at around 20 keV, but are sensitive to beyond 100 keV and, due to the protective layer, down to a couple of keVs. Although the scanners can generally be set to step sizes of as little as 10 μm, the actual spatial resolution achieved is more likely, depending on the model of the scanner and plate, to be in the region of ~100 μm [19]. After the data is extracted, the plate is wiped by exposure to an intense halogen lamp, for some minutes, that clears all remaining trap-level electrons. These plates have been shown to be linear in response over five orders of magnitude [17].

Image plates, of course, have the advantage of not being susceptible to electromagnetic pulse (emp) damage. In addition, they can be large area, typically up to 35 cm × 43 cm, and can be cut to shape. The sensitivity to harder x-rays can be an advantage in some situations but, of course, it also makes them sensitive to hard x-ray background noise in some experiments where we would rather avoid this.

4.2 X-ray scattering

X-ray scattering as a diagnostic of WDM and dense plasmas has been extensively investigated and developed since the 1990s, see for example [11, 20–27]. The technique is at a stage now where sophisticated experiments and analysis can be applied to explore key questions in WDM theory such as plasma collisionality [28] and charge state dynamics in radiatively heated dense plasmas [29].

4.2.1 Scattering models

Following the work of Chihara [30, 31] we can represent x-ray scattering through the following equation;

$$I(k, \omega) = \left(\frac{d\sigma}{d\Omega}\right)_T \Big[(f(k) + q(k))^2 S_{ii}(k, \omega) + Z_b \int S_b(k, \omega - \omega') S_i(k, \omega') d\omega' + Z_f S_{ee}(k, \omega) \Big] \tag{4.5}$$

In this equation, $k = |\mathbf{k}|$, is the magnitude of the scattering wave-vector and the scattering is divided between three principal contributions that are calculated separately. All three contributions are scaled to the classical Thomson scattering cross-section which, for un-polarised light, is given, in SI units, by;

$$\left(\frac{d\sigma}{d\Omega}\right)_T = \left(\frac{e^2}{4\pi\epsilon_0 m_e c^2}\right)^2 \frac{1 + \cos^2\theta}{2} \tag{4.6}$$

where θ is the scattering angle, which is the angle between the directions of the incident and scattered photons and can be related to the scattering wave-vector by;

$$k^2 = (k_s - k_0)^2 + 4k_s k_0 \sin^2(\theta/2) \tag{4.7}$$

where k_0 and k_s represent values of the initial and final wave-vectors for the scattered photon. For elastic scattering, or where the change in photon energy is relatively small, we can re-cast this, using the photon wavelength, λ, as;

$$k = (4\pi/\lambda)\sin(\theta/2) \tag{4.8}$$

With the advent of x-ray free-electron lasers, it is important to consider the case where the x-ray source is polarised, in which case, we need to replace equation (4.6) with;

$$\left(\frac{d\sigma}{d\Omega}\right)_T = \left(\frac{e^2}{4\pi\epsilon_0 m_e c^2}\right)^2 \sin^2\phi \tag{4.9}$$

In this equation, the angle, ϕ is not the same as the scattering angle, but is the angle between the E-field of the incident radiation and the scattering direction.

Elastic scattering contribution

Returning to equation (4.5), the first term inside the square bracket represents the quasi-elastic scattering. Within this term, there are three components. The first, $f(k)$ is the ionic form factor that originates from the bound electrons acting coherently. This term is calculated from integration over the electron density distribution of the atoms or ions in question;

$$f(k) = \int \rho_c(\mathbf{r}) e^{i\mathbf{k}\cdot\mathbf{r}} d^3\mathbf{r} \tag{4.10}$$

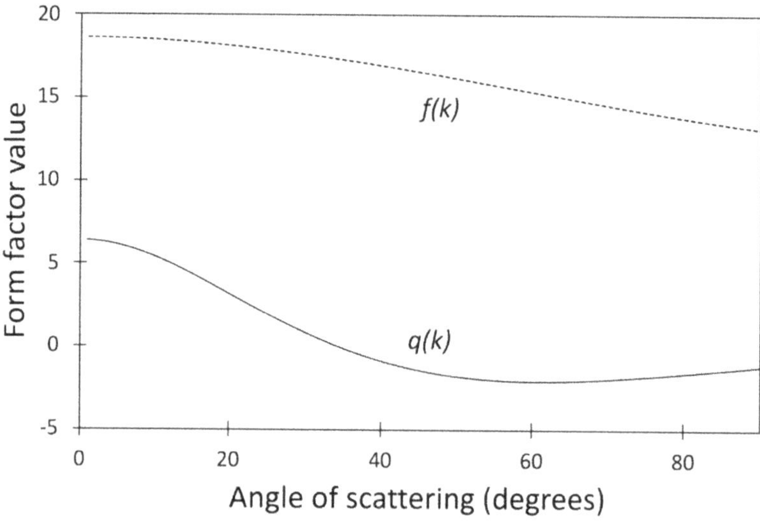

Figure 4.5. Calculation of the ionic-form factor, $f(k)$ and the free-bound electron correlation term, $q(k)$ for Fe under WDM conditions and assuming a photon energy of 7 keV. The latter is calculated with an assumes a potential of the form in equation (4.12), with a cut-off radius of 0.5 Å.

In figure 4.5 we can see an example of such a calculation for the case of shock compressed Fe at a density of 11 g cm^{-3} and temperature 1 eV. The calculation uses the charge density distribution, $\rho_c(\mathbf{r})$ calculated from the Thomas–Fermi Model. At the small scattering angle ($k = 0$) limit, the value of $f(k)$ tends to the number of bound electrons and we see that, in this case, $Z_b \sim 18$. It is the square of the term $f(k)$ that appears in equation (4.5) and so we can see that, for moderately high Z materials, in the WDM state, this will become the dominant contribution to the scattering signal. The term, $q(k)$ accounts for correlation between free electrons and ions and is a little more complex to calculate [32]. It can be represented in terms of partial structure factors, for example;

$$q(k) = \sqrt{\bar{Z}}\frac{S_{ei}(k)}{S_{ii}(k)} \tag{4.11}$$

In this equation, $S_{ii}(k)$ is the ion–ion structure factor that we encountered in chapter 1. The term $S_{ei}(k)$ is a similar term but this time the correlation is between the free electrons and the ions, with average ionisation, $\bar{Z}$. For a partially ionised plasma, the correlation between ions and free electrons is dependent on the free electron interaction with the bound electrons. This can be represented by choosing a potential form to account for this. One simple form that can be chosen is;

$$\begin{aligned} V_{ei}(r) &= \frac{\bar{Z}e^2}{4\pi\epsilon_0 r} \quad r > r_{\text{cut}} \\ V_{ei}(r) &= 0 \quad r < r_{\text{cut}} \end{aligned} \tag{4.12}$$

where we are assuming the bare Coulomb potential beyond some defined distance r_{cut}, with average ion charge $\bar{Z}$. In figure 4.5, we also show $q(k)$ where it is calculated from the work of Gericke *et al* [32]. Although the overall angular distribution of scattered photons is dominated by the ion–ion structure factor, with $f(k)$ varying more slowly and relatively predictably, we should not neglect the effect of $q(k)$ on the observed data. For example, we see in figure 4.5 that $q(k)$ is a significant fraction of $f(k)$ and so will significantly affect the overall scattering as well as the angular behaviour. In addition, we note that, for some angles, $q(k)$ can be negative, indicating destructive interference between scattering from the bound and free electrons. Since $q(k)$ is added to $f(k)$ before squaring, this can lead to narrowing of the angular width of the coherent scattering features. This can be important in analysing the scattering by comparison to simulations in order to estimate the plasma conditions.

The third component of the quasi-elastic scattering, $S_{ii}(k, \omega)$, is the dynamical form of the ion–ion form factor. This is connected to the version of the ion–ion form factor presented in chapter 1 by;

$$S_{ii}(k) = \int S_{ii}(k, \omega)d\omega \tag{4.13}$$

The spectrally integrated version, $S_{ii}(k)$ accounts for the average positioning of the ions whilst the dynamical version accounts for ion motion. The dispersion relation for ion acoustic waves in a plasma is given by;

$$\omega^2 = k^2\frac{\gamma_e k_B T_e + \gamma_i k_B T_i}{M} \tag{4.14}$$

where $\gamma_{e,i}$ are the adiabatic indices for electrons and ions respectively and M is the mass of the ions. If we take a scattering wave-vector of $k \sim 10^{10}\ m^{-1}$ as typical for few-keV photon scattering, then we see that the energy resolution needed is in the 10s of meV and we need a spectral resolution, $\mathrm{E}/\Delta E \sim 10^5$, which the vast majority of x-ray scattering experiments cannot reach. However, although, we do not usually resolve the ion feature, there have been a handful of experiments that have achieved this in the WDM regime. For example, McBride *et al* [33] describe the use of diffraction from a Si crystal to produce a narrow x-ray beam from the broader LCLS beam. Mabey *et al* [34] discuss how such measurements of the dynamic ion–ion structure depend on the detailed transport coefficients and are thus a key to understanding this important aspect of WDM physics.

For calculating both the spectrally integrated and dynamical structure factor, density-functional-theory-molecular dynamics (DFT-MD) simulations are often used as recent advances in computer power enable larger samples to be run for many time-steps. For faster calculations of the spectrally integrated $S_{ii}(k)$, it is common to use a hypernetted-chain (HNC) treatment discussed in chapter 1. An example of fitting to experimental data that makes use of the HNC approximation is shown in figure 4.7.

Bound-free Compton contribution

The second term, within the square brackets of equation (4.5), represents the bound-free Compton scatter. The term, $S_b(k, \omega - \omega')$, represents the bound-free structure factor and $S_i(k, \omega')$ accounts for the thermal motion of the ions. The bound-free term arises when a scattered photon imparts sufficient recoil energy to a bound electron to cause ionisation. As undergraduates, we are introduced to Compton scattering in the case of a free electron at rest. In that case, the energy of the scattered photon is simply related to the angle of scattering by the requirements of energy and momentum conservation. For the case of a bound electron, we do not have a single energy for the scattered photon at any given angle, but a spectrum of energies that depends on the momentum distribution for the wave-function of the initial bound electron. It is for this reason that this technique has also been used to probe the wave-functions of electrons in solids [35]. In the so-called *impulse approximation* [36, 37] which applies for large momentum transfers, the Compton scattering cross-section is given by;

$$\frac{d^2\sigma}{d\Omega d\omega} = \left(\frac{d\sigma}{d\Omega}\right)_T \frac{\omega_s}{\omega_0} \frac{m_e a_B}{\hbar^2 k} J(\xi) \tag{4.15}$$

with ω_0 and ω_s being the incident and scattered photon frequencies respectively and the Compton profile for the scattered photons is given by [35];

$$J(\xi) = 2\pi \int_{|\xi|}^{\infty} p\, |\chi(p)|^2\, d^3p \tag{4.16}$$

where $\chi(p)$ is the initial momentum space wave-function for the bound state and ξ is related to the transferred momentum by;

$$\xi = \frac{m_e a_B}{\hbar k}\left(\omega - \frac{\hbar k^2}{2m_e}\right) \tag{4.17}$$

As above, k is the scattering wave-vector and ω is the frequency shift from the initial photon frequency. Specific analytical contributions for different hydrogenic wave-functions have been calculated [37], for example;

$$J_{1s}(\xi) = \frac{8}{3\pi Z_{\text{eff}}\left(1 + \frac{\xi^2}{Z_{\text{eff}}^2}\right)^3} \tag{4.18}$$

$$J_{2s}(\xi) = \frac{64}{\pi Z_{\text{eff}}}\left[\frac{1}{3\left(1 + 4\frac{\xi^2}{Z_{\text{eff}}^2}\right)^3} - \frac{1}{\left(1 + 4\frac{\xi^2}{Z_{\text{eff}}^2}\right)^4} + \frac{4}{5\left(1 + 4\frac{\xi^2}{Z_{\text{eff}}^2}\right)^5}\right] \tag{4.19}$$

$$J_{2p}(\xi) = \frac{64}{15\pi Z_{\text{eff}}} \frac{\left(1 + 20\frac{\xi^2}{Z_{\text{eff}}^2}\right)}{\left(1 + 4\frac{\xi^2}{Z_{\text{eff}}^2}\right)^5} \tag{4.20}$$

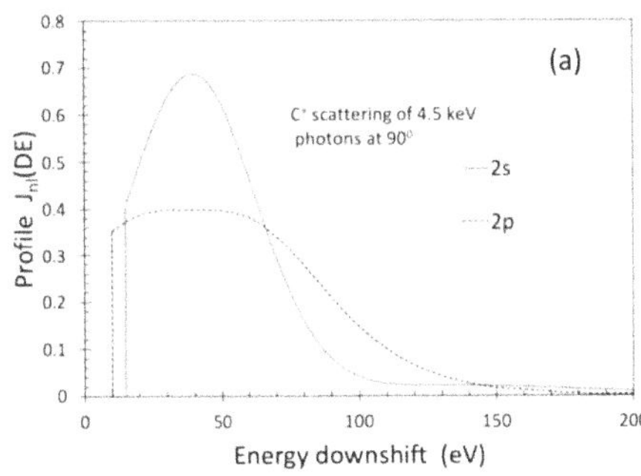

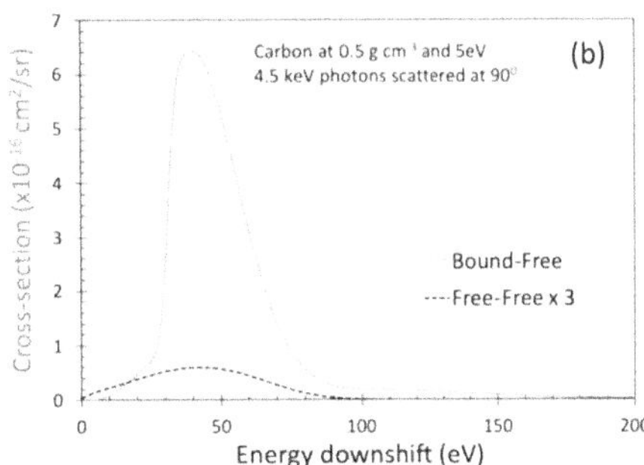

Figure 4.6. (a) Calculation of bound-free Compton profiles using the Bloch-Mendelssohn profiles [37] for singly ionised carbon for 90° scattering of 4.5keV photons. The curves are for one electron only per shell. (b) Calculation made for a carbon WDM sample at 5 eV and 0.5 g cm^3 where there is a mixture of mainly C^+ and C^{2+} ions and 5 eV instrument broadening has been included. As we can see, this process is likely to be dominant over the free-electron Thomson scattering which has been calculated here with a Random Phase Approximation, see discussion below.

where, if Z is the atomic number, then Z_{eff} is an effective atomic number designed to make sure the profile at $\xi = 0$ is the same as would be obtained by using a Hartree–Fock solution to the wave-function rather than a hydrogenic wave-function. This can be calculated using screening constants and depends on the orbital and ionisation stage in question. Profiles for other sub-shells can be found in the literature in several places [26, 37]. In figure 4.6(a), we use these approximations to plot the Compton profile for the case of 4.5 keV photons scattered from the 2s and 2p shells of the C^+ ion. As we can see, the scattering shows a spectrum broad enough that it should be readily measurable experimentally with crystal spectroscopy. The vertical lines indicate where we have forced the scattering profile to drop to zero. We do this because, for lower energy shifts, the electron has not absorbed enough energy from the scattering photons to reach a free state and the process cannot occur. In figure 4.6(b) we have made a calculation for a dense C plasma where there is a mixture of ion stages. Again, we have truncated the profile at the ionisation energy, where the latter includes the effect of ionisation potential depression (IPD) with the Stewart–Pyatt model discussed in chapter 1.

The calculations we have presented in figure 4.6 are made using readily available analytic solutions, and they give a flavour of what the bound-free Compton profile should look like. However, there are some approximations involved that may or may not be accurate. Since the photon energies used in WDM x-ray scattering experiments tend to be in the few keV regime, we expect energy transfers of at most a few hundred eV, and it is important to assess just how accurate impulse approximation treatments can be, especially if we are to use them for determination of WDM conditions, where we do not know *a priori*, the temperature, density and ionisation state.

The impulse approximation should be valid as long as the energy imparted to the bound electron is large compared to its binding energy. This is equivalent to saying that the potential, that we have to consider the electron to be bound in, is the same at the start and end of the collision with the photon. The accuracy is given by $\sim(E_b/E_c)^2$ where E_b is the binding energy and E_c is the energy imparted by Compton scattering. For the 2s orbital in figure 4.6 we can use screening constants [38] to estimate a binding energy around 14 eV, if we account for IPD. At the peak of the Compton profile, ~40 eV, the impulse approximation should be good to roughly 10%. How this translates into the accuracy with which the measured profiles can be used to extract parameters, such as average ionisation from experimental data, will depend in the particular case and some care may be warranted. First order corrections to the impulse approximation profiles were presented by Holm and Ribberfors [39].

Mattern and Seidler [40, 41] have discussed the accuracy of approximations, used in calculation of the bound-free contribution for WDM, that have been presented in the literature, by reference to their performance against cold sample data. They used a real-space Green's function (RSGF) approach that can include the effect of neighbouring ions, even for a disordered system such as WDM. These neighbouring ions can scatter the ejected photo-electrons, leading to possible modulations in scattered spectrum, that might be seen with high resolution experiments. Their calculations were found to be in good agreement with high quality cold sample measurements from synchrotron experiments. They found that, whilst a plane wave form factor approach [36] performed poorly in the same comparison, an impulse approximation with a truncated profile to account for the ionisation potential works reasonably well overall. However, this agreement is tempered by the fact that for individual sub-shells there was less good agreement with the RSGF calculations and that the simple truncation of the profile comes at the expense of not obeying the sum rules for transitions exactly. In summary, although relatively easy to implement expressions are available, on a case by case basis, the applicability of hydrogenic wave-functions and the impulse approximation need to be considered.

Free-electron scattering contribution

The final part of equation (4.5) is the free-electron Thomson scattering. As with optical Thomson scattering, this can occur in either the collective or non-collective modes. There are several treatments of the electron feature for dense plasmas and WDM that have been discussed in the literature [42, 43]. A simple approach is to use the random phase approximation (RPA). This approximation assumes that the various fluctuations in density at different spatial frequencies (the k-vectors) act essentially randomly and are not non-linearly coupled. This approach can give good results in some cases, where collisions are not important, but in many situations it is necessary to go further and include the effects of electron–ion collisions. For example, accounting for the effect of collisions has been attempted, through the use of local field corrections [44]. If we start with the RPA approximation, we can write the dielectric function as;

$$\epsilon_{\mathrm{RPA}}(k, \omega) = 1 - \nu(k)\chi_{\mathrm{RPA}}(k, \omega) \tag{4.21}$$

where $\nu(k) = e^2\epsilon_0 k^2$ is the Fourier transform of the Coulomb potential and $\chi_{\mathrm{RPA}}(k, \omega)$ is the density response function, which in the RPA approximation is:

$$\chi_{\text{RPA}}(k,\omega) = \frac{1}{\hbar}\int_0^{\infty} \frac{f(\mathbf{p}+\hbar\mathbf{k}/2) - f(\mathbf{p}-\hbar\mathbf{k}/2)}{\mathbf{k}\cdot\mathbf{p}/m_e - \omega - i\nu}\frac{2dp^3}{(2\pi\hbar)^3} \tag{4.22}$$

With local field corrections, equation (4.22) becomes;

$$\chi(k,\omega) = \frac{\chi_{\text{RPA}}(k,\omega)}{1 + G(k,\omega)\chi_{\text{RPA}}(k,\omega)\nu(k)} \tag{4.23}$$

The calculation of the local field correction, $G(k, \omega)$, is in general non-trivial and requires approximations that go beyond this text but are discussed in the literature cited in this chapter. In figure 4.7(b), we show an example of collective scattering data, where we see that a plasmon feature is seen on the red-shifted side of the central elastic scattering peak. This asymmetry is typical in the x-ray Thomson scattering regime and is governed by the principle of detailed balance such that there is a ratio of;

$$I(\omega,k) = I(-\omega,k)\exp\left(-\frac{\hbar\omega}{k_BT_e}\right) \tag{4.24}$$

This asymmetry is typically not noticeable in optical Thomson scatter as we generally have $\hbar\omega/k_BT_e \ll 1$ and the exponential term is very close to unity. In WDM this asymmetry is noticeable and is a potential temperature diagnostic that does not depend on the damping of the plasmons or detailed understanding of the plasmon dispersion. In the classical collisionless plasma approximation, the frequency of the plasmon features, for the small scatter wave-vector limit, is given by;

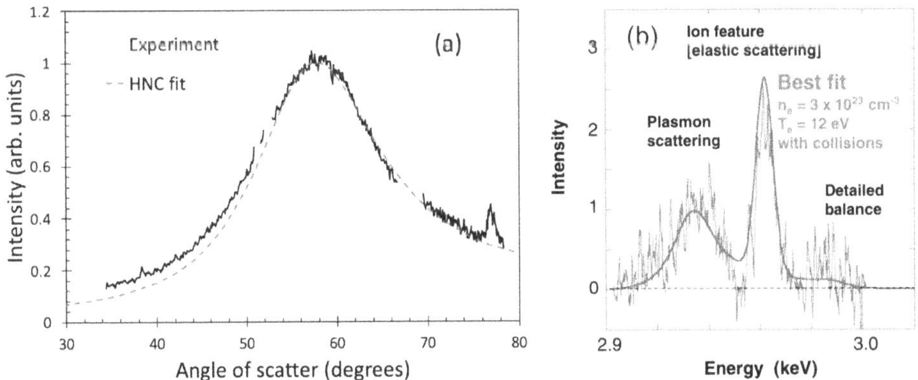

Figure 4.7. (a) Elastic scattering from shock compressed Fe taken with 7 keV photons from the LCLS x-ray free electron laser. The HNC fit is to 11.1 g cm^{-3} and 1.5 eV using a screened ion–ion potential that has a short range component that scales as $\sim 1/r^4$ to simulate bound shell repulsion. For similar data see [51]. (b) X-ray Thomson scattering spectrum taken at 40° with a Be sample, radiatively heated with Ag L-shell radiation using the Cl Ly-α line at 2.96 keV as a probe. We can see the effect of detailed balance in suppressing the plasma feature on the blue side of probe wavelength. Reproduced with permission from [45] © 2007 by the American Physical Society.

$$\omega^2 = \omega_p^2 + \frac{3k_BT_e}{m_e}k^2 \tag{4.25}$$

where, the second term on the right is the Bohm–Gross term and in this long wavelength limit, $\hbar k^2/(2m_e\omega) \ll 1$. For WDM where we have degeneracy and quantum diffraction effects and photon energies that are higher, we must modify this and, as given by Höll *et al* [46, 47] get;

$$\omega^2 = \omega_p^2 + \frac{3k_BT_e}{m_e}k^2(1 + 0.088n_e\Lambda_e^3) + \left(\frac{\hbar k^2}{2m_e}\right)^2 \tag{4.26}$$

for the weakly degenerate ($\mu/k_BT_e < 1$) case. In this equation, $\Lambda_e = h/\sqrt{(2\pi m_e k_B T_e)}$ is the electron thermal de Broglie wavelength and this term accounts for electron degeneracy. The final term is the quantum shift as discussed by Haas *et al* [48]. For typical WDM condition, the modified Bohm–Gross term and the quantum shift term are very important. If we imagine an experiment where we scatter 3 keV photons at 30° for a plasma with $n_e = 2 \times 10^{23}$ cm^{-3} and $T_e = 10$ eV, then we can calculate that the Bohm–Gross term is modified by about 20% due to degeneracy effects and that the effect of both these terms is to shift the peak of the plasmon feature from a low frequency limit of 16.5 eV to around 21 eV.

4.2.2 Some examples of experimental scattering data

There are many good examples of x-ray scattering data in the literature and we limit ourselves, here, to just a couple of examples, one of angularly resolved and one of spectrally resolved data. In figure 4.7(a), we see a normalised scatter cross-section as a function of angle for shock compressed Fe, taken at the MEC (Materials at Extreme Conditions) end-station [49, 50] of the LCLS x-ray laser facility, in this case [51] operating at 7keV photon energy. The scattering is detected directly onto the detectors without spectral dispersion. However, the spectral width of the x-ray beam is less than 1% and so resolves the wave-number of the scattering well. Since we are using a moderately high Z material in the WDM state, the scattering is dominated by the coherent scattering of the first term in equation (4.5).

The experimental data has been fitted using a hypernetted chain approach [52] using a screened Coulombic potential between the ions supplemented by a short-range repulsive term that accounts for the interaction between the shells of bound electrons. As discussed in chapter 1, an important role of such measurements is that it can connect the microscopic structure of the WDM sample to bulk properties such as the thermal and electrical conductivity as well as compressibility and internal energy. As discussed above, figure 4.7(b) shows an example of x-ray Thomson scattering, in the collective mode, showing plasmon features, which can be used as a diagnostic of the sample density. In addition to the ratio of red and blue plasmons, given by equation (4.24), the ratio of the elastic to inelastic scattering has been modelled [53] and used as a diagnostic of the temperature, for similar data.

4.2.3 Sources for x-ray scattering

The scattered signals discussed above are small because they are scaled by the Thomson scattering cross-section, and thus our source of x-rays must deliver as many photons on target as possible and the collection system must, likewise, be efficient. These considerations do depend on whether we are looking at spectrally resolved scattering at a fixed angle or the coherent scattering as a function of angle. For the former, we can make an estimate using a typical value of the Thomson scattering cross-section of $\sim 4 \times 10^{-26}$ cm^2 sr^{-1} for 90° scattering. Now, we can imagine a sample of WDM of order 100 μm thick, with electron density $\sim 10^{23}$ cm^{-3}, which means that a fraction, 4×10^{-5} of incident x-rays will be scattered. We also have to consider that, for spectrally resolved scattering, we need to have a spectrally narrow source that will allow good resolution in the signal.

Highly stripped ions from laser-plasmas

In earlier experiments at large laser facilities, it was common to use K-shell emission lines [54] from He-like and H-like ions, emitted by laser plasmas, as a scattering source. Deciding which materials to use involves a balance between the need to create a larger number of photons and the need to have sufficient photon energy to penetrate the sample without a high degree of absorption. A typical He-α (1s^2-1s2p) line group, for Ti, is shown in figure 4.8(a). We can see that the Li-like satellites form an important part of the line group but that the overall energy spread is approximately 1% of the average photon energy. We have seen above, that the width of the free-electron feature can be wider than these lines and thus, we can still realistically expect to get spectrally resolved data.

The conversion efficiency for laser light to K-shell x-rays has been investigated by several authors, for example [54–56]. In an early classic paper, Phillion and Hailey [54] showed experimental conversion efficiencies for the He-α line group from a range of targets irradiated with 120 ps duration, 527nm laser pulses. In figure 4.8(b) we can see how strongly the conversion efficiency drops with atomic number, from

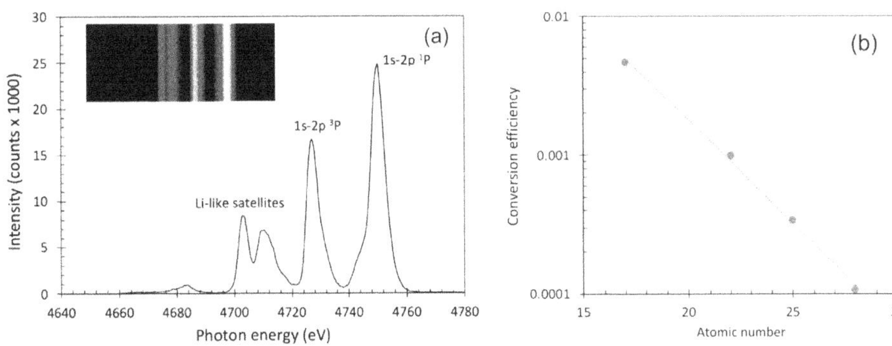

Figure 4.8. (a) Typical He-α spectrum for Ti. The inset shows the raw ccd data. The Li-like satellites mean that the spread of energy in the source is approximately 50 eV. (b) Conversion efficiency of 120 ps, 527 nm laser pulses to He-α line group for data extracted from [54]. The dashed line is an exponential fit to the data points.

close to 0.5% for Cl ($Z = 17$) to barely more than 0.01% for Ni ($Z = 28$). The variation in conversion efficiency closely follows an exponential fit with Z;

$$CE(Z) \sim 1.7e^{-0.34Z} \tag{4.27}$$

For the He-α ($1s^2$-$1s2p\ ^1P$) line, the photon energies range from 2.789 keV for Cl to 7.806 keV for Ni. We can estimate conversion efficiency for materials in between from the exponential fit to the data. We should note that, for this data, the laser intensity at which the conversion efficiency was optimised also varied with the atomic number, ranging from just under 10^{15} W cm^{-2} for Cl to over 10^{16} W cm^{-2} for Ni. The pulse duration of the x-rays was measured to be similar to the 120 ps of the incident laser. For many experiments, this should be short enough to provide sufficient temporal resolution. For longer pulses, the efficiency can be improved, e.g. [55] and although this means a longer x-ray pulse, as we show in figure 4.9, this may still be useful in many experiments e.g. [21]. With a large laser facility, we might expect of order 100 J on target for such pulses. In this case, we can expect to produce between 10^{13} and 10^{15} photons per pulse across this range of elements. In order to achieve what might be considered a usable angular resolution of 0.1 radian, we can achieve up to 10^{11} to 10^{13} photons on target.

With the use of focusing crystals, we can enhance the collection of photons. As noted above, Von-Hamos type crystals with cylindrical focussing can be readily made from HOPG. With its high integrated reflectivity of order 10^{-3} radians and a typical collection angle of ~0.1 radian for the focusing geometry, the effective solid angle for collection can be of order 10^{-4} steradians.

K-α sources

If much shorter duration is desirable, perhaps due to the hydrodynamic timescale of the sample, then we can consider using laser-plasma K-α sources. In these, the interaction of a sub-picosecond pulse, with a solid target, generates fast electrons that penetrate the solid and generate inner shell vacancies that result in K-α emission [57]. The laser duration in such sources is usually picosecond or sub-picosecond, although

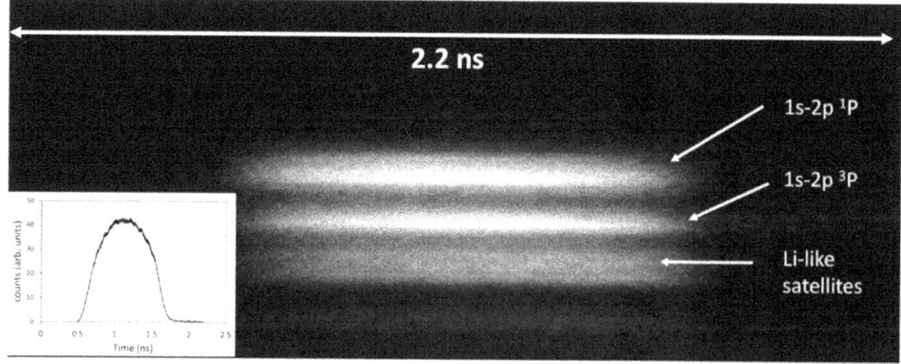

Figure 4.9. X-ray streak of Ti He-α line created with a 1 ns FWHM optical pulse. We can see that the FWHM of the x-ray pulse is similar.

the duration of the x-ray emission can depend on the dynamics of the hot electrons in the target and may be longer than the laser-pulse [58, 59]. The K-α spectrum is a doublet line feature that is relatively narrow, for example, the K-α_1 and K-α_2 for Cu are separated by only 0.38 pm, which is 0.25% of the wavelength. There have been many experimental and theoretical investigations to explore the efficiency of such sources [58, 60]. Typically, conversion efficiencies from laser energy to K-α energy are only of order 10^{-5} to 10^{-4}, although, due to the increased importance of radiative processes, at higher Z, in filling the K-shell vacancy, the efficiency does not drop sharply with Z in the way it does for the He-α source. The low conversion is a disadvantage in x-ray scattering experiments and for this reason it has only been used in a limited number of experiments where a relatively large sample has been available to provide a viable signal, for example [61].

X-ray free electron laser sources
The advent of x-ray free electron sources coupled to powerful optical lasers has revolutionised x-ray scattering from WDM. The source is free from the background of broadband bremsstrahlung and recombination emission associated with laser-plasmas. In addition, the pulse duration is typically < 100 fs, giving excellent temporal resolution which allows experiments to probe processes such as melting and equilibration. The source can be tuned to more or less arbitrary photon energy across the keV range. For example, the upgraded LCLS II facility is expected to be operational over 1–25 keV with photon numbers between 10^{11} and 10^{13} per 1–50 fs pulse on target depending on the energy selected. This is competitive with the yield of He-like emission from laser plasma facilities and can be produced at a high repetition rate with some other key advantages. Firstly, the bandwidth of the beam is typically only a fraction of a percent (typically 0.1–0.3 % for LCLS). Secondly, the divergence of the beam in the keV regime is only a few μradians. In addition, although focussing to sub-μm spots is possible, the size of the unfocussed beam is only a few 10s μm and some experiments may be possible without focussing.

As noted above [33] it is possible to reduce the bandwidth by use of a crystal monochromator. A typical Bragg crystal will reflect the incident radiation with a bandwidth of around $\Delta E/E \sim 10^{-4}$. This narrowing comes at the price of higher shot-to-shot energy fluctuations because the spectrum of the x-ray free electron pulse generally shows a strongly speckled nature, due to the many longitudinal modes present, which is subject to shot to shot variation within the overall bandwidth.

4.3 X-ray absorption measurements

Since WDM samples are often transparent to x-rays, a key diagnostic is absorption spectroscopy. An important focus of experiments is the x-ray absorption in the region around an edge such as a K-edge. As noted in chapter 1, the position of the edge is affected by the dense plasma environment. To recap, this dependence arises, in simple terms, from three factors. The first is a red-shift due to continuum lowering or IPD [62] through which the energy levels of bound electrons are raised by interaction with the ambient plasma electric micro-field. The second term is the blue-shift due to ionisation as a result of the elevated temperature and compression of WDM. Finally, there is the change in degeneracy and the Pauli blocking which, depending on initial and final conditions, can contribute to either a blue or red-shift of the edge. In the first experimental work using this diagnostic for WDM, Bradley *et al* [63] measured a shift of −3.7eV for a KCl sample that was estimated to be at 6.2 g cm^{-3} and 19 eV. This, however, was predicted to be comprised of a continuum lowering of −45.2 eV, an ionisation shift of +47.6 eV and a change due to degeneracy of −6.2 eV. This illustrates a challenge of this diagnostic in that a small net shift is seen and relatively small changes in the theory of IPD or estimates of the ionisation can have a large proportional effect on the final result. The issue of IPD is currently a field of renewed interest since recent experimental data from x-ray laser facilities and large laser facilities disagree on which model is appropriate, see Crowley [64] and references within. As we shall see in the following, the position of the edge is not the only feature of interest. The strong coupling found in WDM gives rise to the short range structure that modifies the absorption of x-rays in the vicinity of the edge.

4.3.1 XANES and EXAFS

A detailed analysis of the spectral region around the edge reveals oscillations in the absorption coefficient. These are caused by quantum mechanical scattering of the ejected free electron from neighbouring atoms. The wave-function of the free electron interferes with the scattered part of the wave-function to create oscillations in amplitude that correspond to the oscillations in the absorption coefficient.

Techniques to look at these features are called EXAFS (extended x-ray absorption fine structure) or XANES (x-ray absorption near edge structure), depending on how far from the edge we are working. These techniques have been widely used in solid state and liquid physics to explore the co-ordination of atoms [65–67] but they can also be applied to dense plasmas and WDM [68–70], where the structure is linked to the microscopic arrangement of the ions. In the single scattering case, where the ejected electron is treated as a plane wave, the oscillation in the absorption coefficient for the sample is described by;

$$\chi(E) = \sum_i \frac{N_i}{kR_i^2} \left|f_i(k)\right| \sin(2kR_i + \phi_i(k)) \exp(-2\sigma_i^2 k^2) \exp\left(-\frac{2R_i}{\lambda_{sc}}\right) \tag{4.28}$$

In this equation, $\chi(E)$ is the oscillatory part of the absorption coefficient defined by;

$$\chi(E) = \frac{\mu(E) - \mu_0(E)}{\Delta\mu_0(E)} \tag{4.29}$$

where $\mu_0(E)$ represents the absorption coefficient of an isolated atom and $\Delta\mu_0(E)$ is the jump in this function at the edge position. The absorption coefficient, for a WDM sample, then depends on the number of ions, N_i within a shell at radius, R_i. It also depends on the amplitude and phase shift of the electron scattering, given by $f_i(k)$ and $\delta(k)$ respectively. The term, σ_i, represents the variance of the positions around the average position, and contains a contribution from the finite temperature, similarly to a Debye-Waller factor in crystal diffraction. The mean free path, λ_{sc}, depends on the wave-number, k, for the photo-electron. Its determination, typically 5–10 Å for cases of interest to us, includes inelastic scattering as well as the effect of the core-hole lifetime, since, in order to participate, the electron must scatter elastically and reach the initial atom before the core-hole created by the x-ray photon is filled, either radiatively or by Auger decay.

The EXAFS technique has been used in shock compressed samples to look at a region relatively far (~100 eV) from the edge. An early example [68] of EXAFS in WDM can be seen in figure 4.10, where a broad band x-ray source from a laser-irradiated uranium target is used to backlight a shock compressed target consisting of Al sandwiched between two CH layers. The density was deduced from the position of the ion–ion correlation absorption peaks and the temperature, which was simulated to reach ~1 eV, has an effect of smoothing out the depth of the oscillations. This effect is more pronounced for higher energies away from the K-edge and will generally limit the use of this technique to the lower temperature end of the WDM regime. In more recent work, using a synchrotron backlighter source, Torchio *et al* [71] reported EXAFS oscillations measurable for Fe shock compressed at 500 GPa with a temperature of ~1.5 eV.

The XANES technique typically looks closer to the edge. The ejected electron has lower energy and multiple scattering from nearby atoms is important. This technique, like EXAFS, is well known in solid state physics but has also been applied to WDM, e.g. [70]. In figure 4.10(b) we see calculations, using *ab initio* molecular dynamics, by Recoules and Mazevet [72], that show the near edge structure for Al as we move from solid (at 900 K) to a melted phase. We see how temperature effects can wash out the structure in the edge as well as broaden it. From figure 4.10, it should be clear that, in order to make the most of the potential of XANES and EXAFS, we should aim to achieve spectral energy resolution, E/ΔE > 500, if not higher, in experiments.

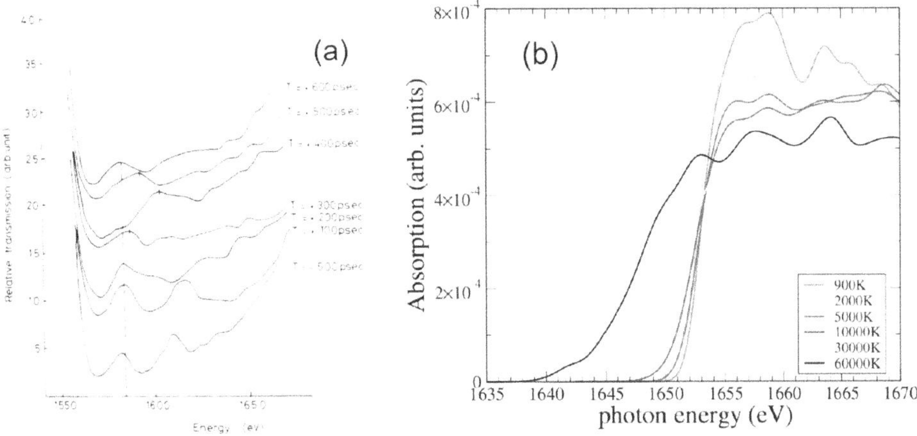

Figure 4.10. (a) EXAFS spectra of a shock compressed Al target, probed at different times in compression in a series of shots. Figure reprinted with permission from [68] © 1988 by the American Physical Society. (b) Simulated XANES profiles for solid density Al at different temperatures. Figure reprinted with permission from [72] © 2009 by the American Physical Society.

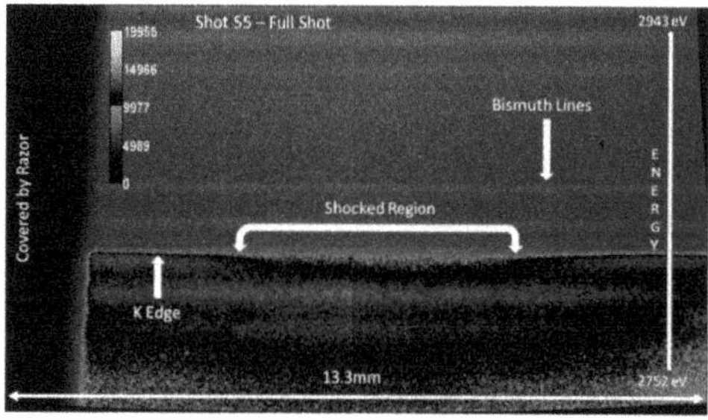

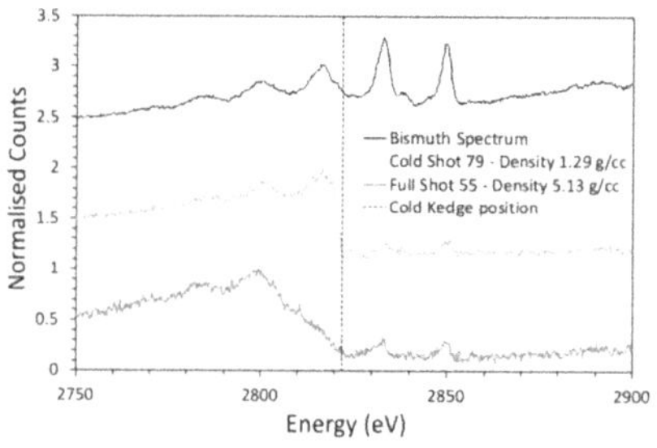

Figure 4.11. (a) Example of background corrected Cl K-edge data. A point source projected through a shock compressed chlorinated plastic sample, allowing spatial resolution across the shocked region in one direction. The colour scheme has been chosen to highlight the area of the shocked region. (b) Line-out of the K-edge data at the shock centre compared to a cold target case. As we see, a clear shift and broadening in the K-edge is visible and measurable for the shocked case. The cold edge position marked is at 2822eV. Images courtesy of David Bailie.

In addition, we need to have a backlighter spectrum that is as smooth as possible in the region of interest, a point we return to in sub-section 4.3.3.

In many experiments with strong shock compression, the temperatures can be of order 10 eV and fine details may be washed out. The focus, then, is often on the position and width of the K-edge. An example of a K-edge measurement for WDM can be seen in figure 4.11 where a Bi M-band source created with a 200 ps duration 527 nm laser pulse focussed at $\sim 10^{15}$ W cm^{-2} has been used to backlight a laser-shock compressed sample containing chlorinated plastic (C-parylene). We can see the K-edge has shifted significantly compared to a cold target. The small source broadening following from the small focal spot possible for the Bi source (< 100 μm) has allowed resolution of 1.4 eV to be achieved. We can notice that there are some well identified Bi features that, in principle, can be used as a spectral fiducial in addition to the cold edge in the spatially resolved data. However, some care must be taken. In this experiment, this would lead to an erroneous calculation of the cold edge position. This is because the Bi source is a laser plasma and hydrodynamic analysis indicated that the most emitting part of the plasma is expanding rapidly at velocities of a few times 10^5 ms^{-1}, leading to a red-shift of the spectrum of several eV.

There have been several attempts to model and measure the shift in K-edge as a function of shock compression, for example [63, 73]. Alkuzee *et al* [74] used a modified Stewart–Pyatt model with a TF average atom model giving the average ionisation but where the exponential function describing the radial distribution of ions is replaced with a radial distribution function, $g(r)$, calculated from a hyper-netted-chain solution to the OCP model. For aluminium, shock compressed up to twice solid density, they found a good agreement for the width and shift of the edge, which showed a maximum of around −5 eV. Zhao *et al* [75] have seen good agreement for compressed KCl with the model used by Bradley *et al* [63] based on the ion sphere model for IPD. It is worth noting, here, that an additional complication in interpreting the K-edge measurements, is the role of line absorption. At high density, orbitals may undergo significant Stark broadening and may merge into one another [76–78]. An orbital close to the ionisation edge may broaden and merge with the continuum, for example the chlorine 3p orbital in shock compressed KCl, as discussed by Zhao *et al* [79]. This can make interpretation of the edge position, in terms of an IPD model, more difficult.

The advances in computing power and speed over the years means that *ab initio* molecular dynamics simulations are now routinely possible with temperatures in the regime ~10 eV that are created in strong shocks. Mazevet *et al* [80] have used such calculations that show good agreement with data, with little dependence of the shift on temperature up to around 1eV but show a sharp increase in shift at higher compressions. Comparisons of *ab initio* molecular dynamics with experiments at higher compressions [81] showed a divergence in agreement at higher than twice solid density.

4.3.2 X-ray line absorption

We have noted above that Stark broadening can lead to merging of bound states, with each other and the continuum. This can lead to the disappearance of sharp absorption lines, that can be seen at lower density. For example, in solid Al, the $3s^2$ and 3p electrons are in the conduction band and the 1s-3p absorption line is not seen. As the density of an Al WDM sample drops, the absorption line may appear and the width of the profile may be compared to Stark broadening calculations. An example of such an experiment is given by Lecherbourg *et al* [82], who irradiated thin foils of Al coated onto a Si_3N_4 substrate, with a 0.3 ps duration pulse laser. A second, synchronised pulse was used to create a broad band, short pulse, x-ray source, by focussing onto a samarium target at $\sim 10^{15}$ W cm^{-2}. They observed both 1s-2p and 1s-3p absorption lines by probing picoseconds after heating, when the foils where at ~0.1 g cm^{-3} density, a few eV in temperature and the strong coupling parameter, $\Gamma > 10$.

We can get an estimate of the likely density at which a line will become bound by carrying out a simple solution to the Schrödinger equation for an atom confined in a box and varying the size of the box, assuming the wave-function goes to zero at the boundaries. For example, in figure 4.12 we see the results of such a calculation for Al. As, we can see, the 3s and 3p orbitals are predicted to have positive energy for density at or above the normal solid value. However, as the box size increases, we can see that the 3s and 3p orbitals take on negative energy and are thus bound. For Al, this occurs at a density of around 1 g cm^{-3}. At this density, we would expect significant broadening of the orbitals, which would may make observation of the 1s-3p transition difficult. However, at lower density, the width of the lines can, in principle, act as a density diagnostic, either if a sample is probed at the lower density, or as a way of providing some validation of hydrodynamic modelling of a sample created and probed at higher density.

4.3.3 Sources for x-ray absorption

An important experimental detail for any of the absorption techniques is the need for a relatively smooth backlight spectrum. This is needed so that it is easy to reliably process out the source spectrum in order to measure the fluctuations in absorption coefficient of the WDM sample.

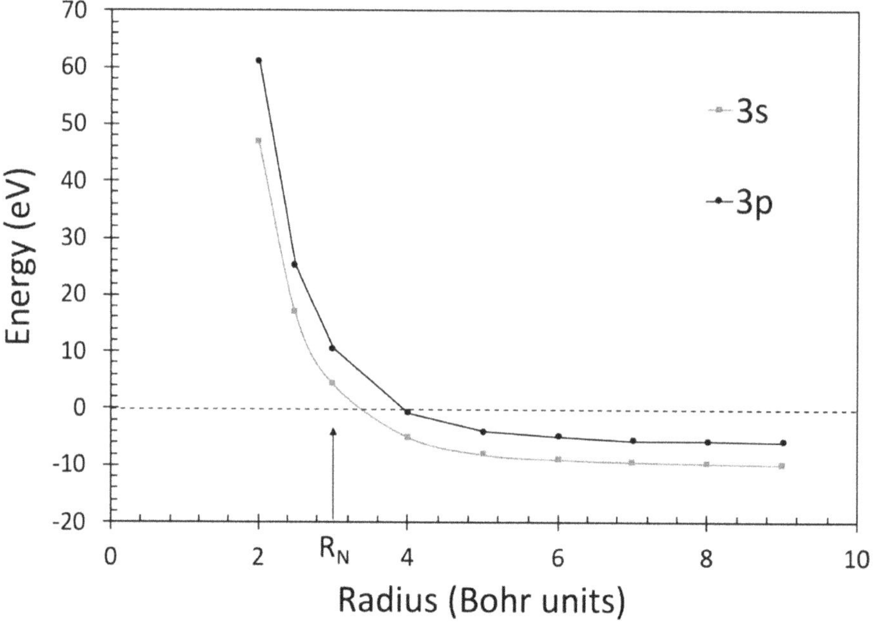

Figure 4.12. Results of solutions to the Schrödinger for an Al atom confined to a box. The point, R_N, indicates the box size for an atom in a solid Al sample. Calculations courtesy of David Waide.

Laser-plasma sources

As discussed above, bremsstrahlung and recombination radiation, from a laser-plasma, should provide a smooth spectrum. However, in order to have a bright enough spectrum, typically, a laser plasma source is used with a higher Z target so that arrays of unresolved transitions form bands of emission. Commonly the M-band of transitions is used. An example spectrum for Au was shown in chapter 3. We see that over a keV range, the spectrum does not look very smooth. However, within the limited spectral range of interest in K-edge shift experiments and to an extent, XANES measurements, the spectrum can be smooth enough to be used successfully. This was seen in figure 4.11(b). We can see that although, there are some clear line features, we do have a broad spectrum, smooth enough for measurements of the K-edge shift and width.

For EXAFS, we are interested in a wider spectral range and, in the pioneering work of Hall *et al* [68] a smoother N and O band source created with a U target was used. The results can be seen in figure 4.10, where we clearly see the oscillations in the transmission of the source through the shock compressed Al sample.

Betatron and synchrotron sources

Over the last decade there has been a considerable body of work on developing compact electron acceleration though mechanisms such as laser-wakefield acceleration (LWFA). An important by-product of this work is the development of so-called betatron sources of x-rays. This occurs when electrons are trapped in the potential created by an intense, relativistic, short pulse laser in a gas target and oscillated side-to-side by the strong electric field as they are accelerated in a forward direction by the ponderomotive force of the laser. The electrons are oscillated at the betatron frequency given by;

$$\omega_\beta = \frac{\omega_p}{2\gamma} \tag{4.30}$$

where ω_p is the plasma frequency, which for a typical electron density of $\sim 10^{19}$ cm^{-3} in LWFA, is around 1.8×10^{14} rad^{-1}. The Lorentz factor, γ is typically >200 and the wavelength of emission is of the order;

$$\lambda = \frac{\lambda_\beta}{\gamma^2} \tag{4.31}$$

where $\lambda_\beta = 2\pi c/\omega_\beta$. This process has some similarity to the wiggler used in conventional synchrotron technology. The resulting strong broadband synchrotron radiation is peaked in the direction of travel. A key advantage of this process is that the source size is only of $\sim\mu$m size, thus allowing for high spatial and spectral resolution without significant source broadening.

Roughly typical laser conditions for LWFA are approximately 50 fs duration, 800 nm wavelength pulse and 10 J focussed with large f-number to several times 10^{18} W cm^{-2}. For this we can expect of order 10^9 photons/pulse with a peak at a few 10s keV. The x-ray are highly directional with a divergence typically <100 mrad and a peak brightness of order 10^{19} photons/s/mrad^2/mm^2/0.1 % bandwidth. In figure 4.13 we can see an actual spectrum from an experiment by Behm *et al* [83]. As we can see, even with single shot data, we can produce a relatively smooth spectrum in a range suited to edge absorption spectroscopy of low to mid-Z elements. Mahieu *et al* [85] have used such a source to make x-ray absorption edge measurements of warm dense Cu created with a femtosecond optical pulse.

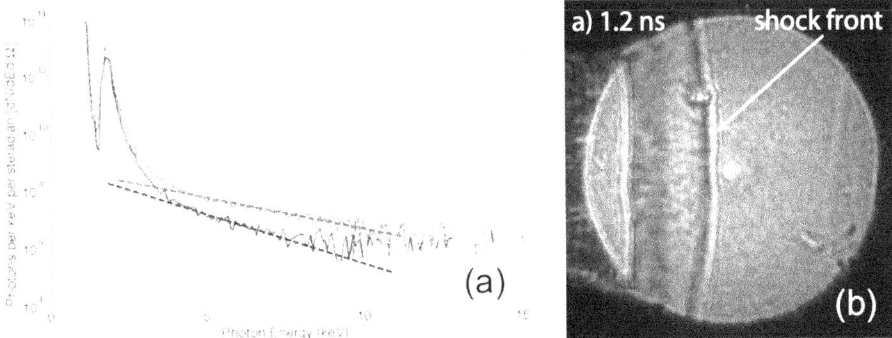

Figure 4.13. (a) Betatron spectrum reproduced with permission from [83], © IOP Publishing Ltd. CC BY 3.0, using a 3J, 35 fs Ti:Sapphire laser operating at 800nm wavelength. The spectrum has been corrected for filtering and the quantum efficiency of the CCD detector used. The red and blue curves refer to different experimental configurations. (b) Phase contrast image of a shock wave taken with an x-ray free electron laser. Image reproduced with permission from [84], CC BY 4.0. For scale, the width of the image is approximately 150 μm.

We noted above, that the work of Torchio *et al* [71] was carried out using a synchrotron source. This is not yet commonplace, as a means of creating a WDM sample, such as a high power pulse laser, needs to be installed and synchronised with the synchrotron beam, but it has become possible since the latest generation of synchrotron facilities are capable of bright pulses of x-rays even down to the femtosecond level [86]. Torchio *et al* used pulses of ~100 ps in duration, and this maintained sufficient photon numbers for single shot measurements, which, for many WDM experiments, provides a short enough temporal resolution to be useful.

4.4 X-ray phase contrast imaging

The small source size and short duration of betatron radiation discussed above, also means that it is suited to carrying out phase contrast imaging of WDM samples. This is an evolving technique that has used both laser-plasma thermal x-ray sources and free electron lasers, with a key application, in the WDM field, being the observation of shock fronts [84, 87]. The technique exploits the change in phase of x-rays passing through an object. The refractive index of the x-rays can be written as;

$$n_r = 1 - \delta + i\beta \tag{4.32}$$

where β represents the absorption and δ is a change in the real refractive index that can be approximated [88], for a wavelength, λ, away from an edge, by;

$$\delta(\lambda) = n_e \frac{r_0 \lambda^2}{2\pi} \tag{4.33}$$

where r_0 is the classical electron radius. For WDM experiments, the free space propagation method is the easiest to implement. In this, a small x-ray source illuminates the target, which is placed between the source and a detector. Changes in refractive index refract the x-rays and, at the image plane, changes in the phase of x-ray coming from different parts of the object, lead to amplitude modulations that reflect the structure, especially for sharp gradients. Phase contrast imaging can be more sensitive to small changes in density than x-ray radiography, which depends on the absorption, which can be weak for lower Z materials. Endrizzi (see [88] and references within) gives an expression for the intensity at a detector surface, for a given wavelength and distances;

$$I(Mx, My) = \frac{I_0}{M^2}\left[1 + \frac{R_{od}\lambda}{2\pi M}\nabla_\perp^2 \phi(x, y)\right] \tag{4.34}$$

where, the magnification, $M = (R_{so} + R_{od})/R_{so}$ and R_{so} and R_{od} are the source to object and object to detector distances, respectively. This equation assumes a phase object without any absorption effects. However, it does show us how important strong gradients in the phase are, and this helps explain why it has been successfully applied to shock wave studies.

The scale-length of features to be imaged has to be smaller than the transverse coherence length. If the distance between source and sample is R_0 and then wavelength of the x-rays is λ, then the transverse coherence length, l_c, in the sample, in the Fresnel diffraction limit is given by;

$$l_c \approx \frac{R_0\lambda}{s} \tag{4.35}$$

where s is the source size. Thus, we can see that, if we take a typical wavelength of 2 Å and the sample to source as ~20 cm, then, in order to have a transverse coherence length of 10 μm, we need to have a source size of only 4 μm. This size of x-ray source is difficult to produce with laser plasmas, but has been done for one dimension, with wire targets, for example [87]. Although strictly monochromatic sources are not absolutely necessary for this technique to be used [89], the bandwidth of the x-ray source will play a role in reducing the contrast of images, with different coherence lengths for different wavelengths.

For an x-ray free electron laser, the source size might typically be in the range 10–20 μm in the keV regime. However, the very high collimation means that focusing of the beam to ~0.1 μm is possible [84], allowing for high quality images, as we can see in figure 4.13(b), where a phase contrast image of a shock propagating in diamond is shown. With x-ray FELs capable of highly monochromatic beams, focusable to very small spot sizes, this technique may become a much more widely used tool.

4.5 X-ray emission spectroscopy

Although we have stated above that emission spectroscopy is of limited value in WDM experiments, there are occasions when it can be used. For example, we have discussed the fact that fast electron heating can be used to create bulk heating of a solid due to colder, more resistive, electrons in the return current. During this process, the fast electrons can collisionally ionise the inner-shell electrons of the sample. The subsequent K-α emission will come from a range of ionisation states depending on the temperature in the sample. We should also acknowledge that, by its nature, this diagnostic is time integrating as the K-α emission will occur as soon as, and for as long as, the fast electrons penetrate the sample. The K-α lines from adjacent ionisation stages are close in energy and high-resolution spectroscopy with $E/\Delta E > 2000$ is desirable. In addition, since we are at high density and Stark broadening may broaden the lines into each other, detailed modelling capability [90] is desirable in order to extract the information on plasma conditions.

In figure 4.14 we see an example of such data using a Ti foil sample. The foil is irradiated at $>10^{18}$ W cm^{-2} and subsequent analysis of similar data [91] suggested fast electrons generated via a resonance absorption process with temperatures of order 50–100 keV. Analysis of the K-α spectrum created by the fast electrons suggested an ionisation spread consistent with a background temperature of up to 20 eV. Even with spatially and spectrally resolving crystals, the small size of the plasma created makes it difficult to separate out contributions from regions under different conditions. In this case, the spectrum was best fitted by assuming emission coming from a 20 eV sample with some contribution from cold Ti. For thicker foils it was found that the spectrum could be modelled by assuming an increasing fraction of the

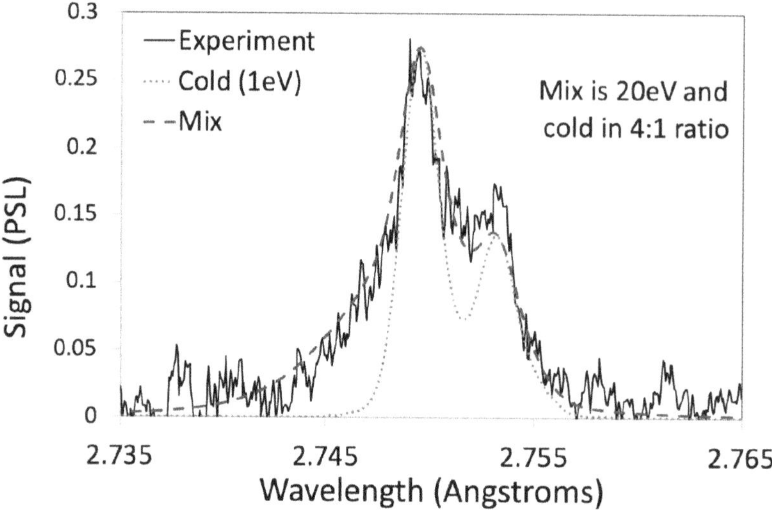

Figure 4.14. Ti K-α spectrum from a 10 μm foil irradiated with a sub-ps, 1.053 μm wavelength pulse at $>10^{18}$ W cm^{-2}. The signal level on the image plate used to record the spectra are in photo-stimulated units (PSU). The fit curves are from the SCRAM code [90] for different assumed background temperatures. Image reprinted from [91], with the permission of AIP publishing ©2014 AIP.

signal came from colder material. In fact, the 'cold' material was modelled as being at 1 eV but showed very little difference from the spectrum expected for Ti at ambient temperature. For Ti, there is only a small energy shift of less than 2 eV between the K-α lines from the first five ion stages and so, even at temperatures above 1 eV, the main effect of heating is to fill in the gap between the K-α_1 and K-α_2 peaks. As temperatures of 10 eV or more are reached we start to see strong contributions from higher ion stages at shorter wavelengths. Depending on the intensity used, spectral resolution achieved, and the element studied, this implies a practical limit to the thickness of target as fast electrons travelling further in than the resistively heated region may contribute a strong cold component to the spectrum, limiting its use in determining the WDM temperature. Chen *et al* [92] were able to use this technique of WDM creation and diagnosis with few micron thickness, mass limited, targets, in which the fast electrons can recirculate as they reach the boundary of the foil to allow for more uniform heating. In this way, they created solid density Ti at bulk temperatures up to ~100 eV at close to solid densities. Detailed modelling was still required for interpretation as, in this experiment, there was also a small hot dense plasma region with T_e> 1 keV in the region of the laser-foil interaction that contributed to the spectrum.

In the example above, the fast electrons are a source of sample heating but also the means by which emission from such a relatively cold sample can occur in the x-ray regime, allowing probing of the bulk conditions. Another interesting example of fast electrons being used takes us to temperatures a bit higher than the usual WDM regime but it is worth including here as the experiment dealt with issues of relevance for WDM. Hoarty *et al* [93] used long pulse laser pulses to shock compress a sample of Al, sandwiched between layers of plastic, to densities up to 10 g cm^{-3}. Under these conditions we do not expect x-ray emission. However, they then heated the sample with a powerful short pulse beam (100 J at 2ω and focussed to $\sim 10^{19}$ W cm^{-2}). This heated the sample to over 500 eV and generated He-like emission from the Al ions. Stark broadening of these lines was used to infer the density and this could be compared to the series cut-off due to the IPD. Their finding was that the Stewart–Pyatt model fitted their data well. Although the sample was relatively hot and thus not very degenerate, the ion–ion coupling parameter $\Gamma \sim 3$ assuming equal electron and ion temperatures and thus at least one criterion for WDM is met.

It is worth pointing out, for these last two examples, that the diagnostic is itself intimately related to the WDM creation process. In the first case, the fast electrons are central to both the resistive heating process and the creation of inner shell radiation. In the second, the heating of the plasma achieved by the fast electrons is important in the determination of the IPD that is diagnosed by the K-shell emission that it allows. This type of situation is not so uncommon and, to some extent it shows that our division of this book into chapters on creation and diagnosis of WDM is simply one of several choices that are possible.

References

[1] Bragg W H and Bragg W L 1913 *Proc. R. Soc. Lond.* A **88** 428–38

[2] Kestenbaum H L 1973 *Appl. Spectrosc.* **27** 454–6

[3] Sanchez del Rio M, Gambaccini M, Pareschi G, Taibi A, Tuffanelli A and Freund A 1998 *Proc. SPIE* **3448** 246

[4] Legall H, Stiel H, Arkadiev V and Bjeoumikhov A A 2006 *Opt. Express* **14** 4570–6

[5] Henke B L, Gullikson E M and Davis J C 1993 *At. Data Nucl. Data Tables* **54** 181–342

[6] Förster E, Gabel K and Uschmann I 1992 *Rev. Sci. Instrum.* **63** 512–6

[7] Monot P, Auguste T, Dobosz S, D'Oliveira P, Hulin S, Bougeard M, Faenov A Y, Pikuz T A and Skobelev Y 2002 *Nucl. Instrum. Methods Phys. Res.* A **484** 299–311

[8] Rosmej F B, Lisitsa V S, Schott R, Dalimier E, Riley D, Delserieys A, Renner O and Krousky E 2006 *Europhys. Lett.* **76** 815–21

[9] Holst Gerald C and Lomheim Terence S 2011 *CMOS/CCD Sensors and Camera Systems 2nd Edition* (Bellingham, WA: SPIE)

[10] Kraft R P, Nousek J A, Lumb D H, Burrows D N, Skinner M A and Garmire G P 1995 *Nucl. Instrum. Methods Phys. Res.* A **366** 192–202

[11] Riley D, Woolsey N C, McSherry D, Weaver I, Djaoui A and Nardi E 2000 *Phys. Rev. Lett.* **84** 1704–7

[12] Stoekl C *et al* 2006 *Rev. Sci. Instrum.* **77** 10F506

[13] Harada T, Teranishi N, Watanabe T, Zhou Q, Bogaerts J and Wang X 2020 *Appl. Phys. Express* **13** 016502

[14] Koerner L J, Philipp H T, Hromalik M S, Tate M W and Gruner S M 2009 *J. Instrum.* **4** P03001

[15] Herrmann S *et al* 2013 *Nucl. Instrum. Methods Phys. Res.* A **718** 550–3

[16] Blaj G *et al* 2015 *J. Synchrotron Radiat.* **22** 577–83

[17] Gales S G and Bentley C D 2004 *Rev. Sci. Instrum.* **75** 4001–3

[18] Meadowcroft A L, Bentley C D and Stott E N 2008 *Rev. Sci. Instrum.* **79** 113102

[19] Fiksel G, Marshall F J, Mileham C and Stoeckl C 2012 *Rev. Sci. Instrum.* **83** 086103
[20] Nardi E 1991 *Phys. Rev.* A **43** 1977–82
[21] Glenzer S H, Gregori G, Lee R W, Rogers F J, Pollaine S W and Landen O L 2003 *Phys. Rev. Lett.* **90** 175002
[22] Woolsey N C, Riley D and Nardi E 1998 *Rev. Sci. Instrum.* **69** 418–24
[23] Ma T *et al* 2013 *Phys. Rev. Lett.* **110** 065001
[24] Glenzer S H and Redmer R 2009 *Rev. Mod. Phys.* **81** 1625
[25] Gregori G, Glenzer S H, Rozmus W, Lee R W and Landen O L 2003 *Phys. Rev.* E **67** 026412
[26] Gregori G *et al* 2004 *Phys. Plasmas* **11** 2754
[27] Glenzer S H *et al* 2010 *High Energy Density Phys.* **6** 1–8
[28] Neumayer P 2010 others *Phys. Rev. Lett.* **105** 075003
[29] Gregori G *et al* 2008 *Phys. Rev. Lett.* **101** 045003
[30] Chihara J 1987 *J. Phys. F: Met. Phys.* **17** 295–304
[31] Chihara J 2000 *J. Phys.: Condens. Matter* **12** 231
[32] Gericke D O, Vorberger J, Wunsch K and Gregori G 2010 *Phys. Rev.* E **81** 065401
[33] McBride E E *et al* 2018 *Rev. Sci. Instrum.* **89** 10F104
[34] Mabey P, Richardson S, White T G, Fletcher L B, Glenzer S H, Hartley N J, Vorberger J, Gericke D O and Gregori G 2017 *Nat. Commun.* **8** 14125
[35] Williams B E 1977 *Compton Scattering* (New York: McGraw-Hill)
[36] Schumacher M, Smend F and Borchert I 1975 *J. Phys.* B **8** 1428–39
[37] Bloch B J and Mendelsohn L B 1974 *Phys. Rev.* A **9** 129–55
[38] Faussurier G, Blancard C and Renaudin P 2008 *High Energy Density Phys.* **4** 114–23
[39] Holm P and Ribberfors R 1989 *Phys. Rev.* A **40** 6251–9
[40] Mattern B A, Seidler G T, Kas J J, Pacold J I and Rehr J J 2012 *Phys. Rev.* B **85** 115135
[41] Mattern B A and Seidler G T 2013 *Phys. Plasmas* **20** 022706
[42] Höll A, Redmer R, Ropke G and Reinholz H 2004 *Eur. Phys. J.* D **29** 159–62
[43] Redmer R, Reinholz H, Ropke G, Thiele R and Höll A 2005 *IEEE Trans. Plasma Sci.* **33** 77–84
[44] Fortmann C, Wierling A and Röpke G 2010 *Phys. Rev.* E **81** 026405
[45] Glenzer S H *et al* 2007 *Phys. Rev. Lett.* **98** 065002
[46] Höll A *et al* 2007 *High Energy Density Phys.* **3** 120–30
[47] Thiele R, Bornath T, Fortmann C, Höll A, Redmer R, Reinholz H, Röpke G and Wierling A 2008 *Phys. Rev.* E **71** 026411
[48] Haas F, Manfredi G and Feix M 2000 *Phys. Rev.* E **62** 2763–72
[49] Nagler B *et al* 2015 *J. Synchrotron Radiat.* **22** 520
[50] Glenzer S H *et al* 2016 *J. Phys.* B **49** 092001
[51] White S *et al* 2020 *Phys. Rev. Res.* **2** 033366
[52] Wünsch K, Vorberger J and Gericke D O 2009 *Phys. Rev.* E **79** 010201
[53] Fletcher L B *et al* 2015 *Nat. Photon.* **9** 274–9
[54] Phillion D W and Hailey C J 1986 *Phys. Rev.* A **34** 4886
[55] Riley D, Woolsey N C, McSherry D, Khattak F Y and Weaver I 2002 *Plasma Sources Sci. Technol.* **11** 484–91
[56] Urry M K, Gregori G, Landen O L, Pak A and Glenzer S H 2006 *J. Quant. Spectrosc. and Radiat. Transfer* **99** 636–48
[57] Chen H, Soom B, Yaakobi B, Uchida S and Meyerhofer D D 1993 *Phys. Rev. Lett.* **70** 3431–4
[58] Salzmann D, Reich C, Uschmann I, Förster E and Gibbon P 2002 *Phys. Rev.* E **65** 036402

[59] Riley D *et al* 2006 *J. Quant. Spectrosc. Radiat. Transfer* **99** 537–47
[60] Rousse A, Audebert P, Geindre J P, Fallies F, Gauthier J C, Mysyrowicz A, Grillon G and Antonetti A 1994 *Phys. Rev.* E **50** 2200–7
[61] Kritcher A L *et al* 2009 *Phys. Rev. Lett.* **103** 245004
[62] Stewart J C and Pyatt K D 1966 *Astrophys. J.* **144** 1203
[63] Bradley D K, Kilkenny J, Rose S J and Hares J D 1987 *Phys. Rev. Lett.* **59** 2995–8
[64] Crowley B J B 2014 *High Energy Density Phys.* **13** 84–102
[65] Norman D 1986 *J. Phys. C: Solid State Phys.* **19** 3273
[66] Koningsberger E and Prins D 1987 *X-ray absorption: Principles, Applications, Techniques of EXAFS, SEXAFS and XANES* (New York: Wiley Interscience)
[67] Newville M 2014 *Rev. Mineral. Geochem.* **78** 33–74
[68] Hall T A, Djaoui A, Eason R W, Jackson C L, Shiwai B, Rose S J, Cole A and Apte P 1988 *Phys. Rev. Lett.* **60** 2034–7
[69] Gordon F I 1993 *Plasma Phys. Control. Fusion* **35** 1207–14
[70] Lévy A *et al* 2009 *Plasma Phys. Control. Fusion* **51** 124021
[71] Torchio R *et al* 2016 *Sci. Rep.* **6** 26402
[72] Recoules V and Mazevet S 2009 *Phys. Rev.* B **80** 064110
[73] Riley D, Willi O, Rose S J and Afshar-Rad T 1989 *Europhys. Lett.* **10** 135–40
[74] Al-Kuzee J, Hall T A and Fry H D 1998 *Phys. Rev.* E **57** 7060–5
[75] Yang Z *et al* 2013 *Phys. Rev. Lett.* **111** 155003
[76] Salzmann D 1998 *Atomic Physics in Hot Plasmas* (Oxford: Oxford University Press)
[77] Griem H R 1964 *Plasma Spectroscopy* (New York: McGraw-Hill)
[78] Inglis D R and Teller E 1939 *Astrophys. J.* **90** 439
[79] Zhao S, Zhang S, Kang W, Li Z, Zhang P and He X-T 2015 *Phys. Plasmas* **22** 062707
[80] Mazevet S and Zérah G 2008 *Phys. Rev. Lett.* **101** 155001
[81] Benuzzi-Mounaix A *et al* 2011 *Phys. Rev. Lett.* **107** 165006
[82] Lecherbourg L, Renaudin P, Bastiani-Ceccotti S, Geindre J-P, Blancard C, Cossé P, Faussurier G, Shepherd R and Audebert P 2007 *High Energy Density Phys.* **3** 175–80
[83] Behm K T *et al* 2016 *Plasma Phys. Control. Fusion* **58** 055012
[84] Schropp A *et al* 2015 *Sci. Rep.* **5** 11089
[85] Mahieu B, Jourdain N, Phuoc K T, Dorchies F, Goddet J-P, Lifschitz A, Renaudin P and Lecherbourg L 2018 *Nat. Comm.* **9** 3276
[86] Schoenlein R W, Chattopadhyay S, Chong H H W, Glover T E, Heimann P A, Shank C V, Zholents A A and Zolotorev M S 2000 *Science* **287** 2237–40
[87] Antonelli L *et al* 2019 *Europhys. Lett.* **125** 35002
[88] Endrizzi M 2018 *Nucl. Instrum. Methods Phys. Res.* A **878** 88–98
[89] Wilkins S W, Gureyev T E, Gao D, Pogany A and Stevenson A W 1996 *Nature* **384** 335–8
[90] Hansen S B *et al* 2005 *Phys. Rev. E.* **72** 036408
[91] Makita M *et al* 2014 *Phys. Plasmas* **21** 023113
[92] Chen S N *et al* 2007 *Phys. Plasmas* **14** 102701
[93] Hoarty D J *et al* 2013 *Phys. Rev. Lett.* **110** 265003

Chapter 5

Optical diagnostics

In warm dense matter experiments, the electron density is much higher than the critical density for optical radiation and so optical probing of the interior is excluded. This also means that optical emission spectroscopy cannot be used to give a clue as to conditions within a sample. Nevertheless, there are several techniques that have been applied that make use of optical radiation in the study of WDM, either in emission or as a probe. For some of these techniques, we have to rely on the assertion that we can relate the surface conditions to the bulk properties. For others, we can use optical probes to diagnose the speed of a shock-wave or of a sample surface without worrying so much about whether the surface conditions are an exact match to the bulk. Before we start to discuss optical diagnostic techniques, we will start with a short outline of optical streak cameras, which have been a workhorse of fast optical diagnostics in plasmas physics over the past 50 years, see for example [1, 2], and latterly in WDM research.

5.1 Streak cameras

In figure 5.1, we can see the basic working principles of the optical streak tube. Light enters a slit and is imaged through the tube window and onto a semi-transparent photocathode. The photons are absorbed and release photoelectrons which are accelerated by a mesh, just behind the photocathode. The applied field is of order 30 KV/cm. After this, a further potential of typically 10–15 KV accelerates the electrons towards a phosphor. Within the tube, an applied radial voltage acts as an electrostatic lens to image the pattern of electrons coming from the photocathode onto the phosphor. The image of the slit on the photocathode is thus imaged onto the phosphor. Application of a fast rising KV pulse, across the sweep plates within the tube, means the image of the slit is swept across the phosphor. As the light pulse, falling onto the photocathode, rises and falls with time, the electron current coming off it does so too. As the electrons hit the phosphor, their kinetic energy is converted to optical emission and because of the sweep voltage, the rise and fall in time of the signal is transformed into a rise and fall of optical emission spatially across the

doi:10.1088/978-0-7503-2348-2ch5

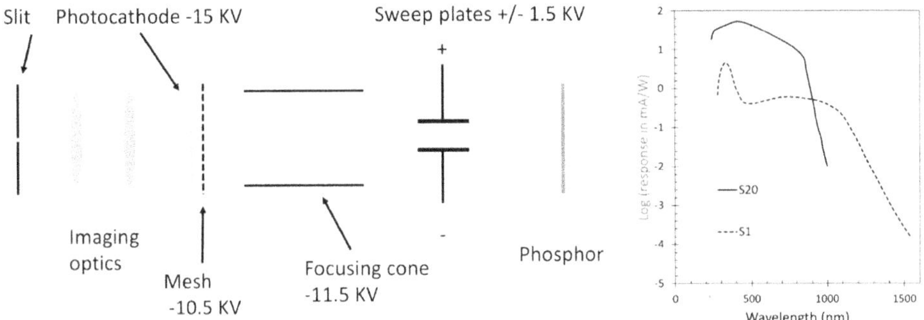

Figure 5.1. Left: a simplified schematic of a streak tube. Right: typical response curves for a streak tube. The S1 and S20 are two common photocathode types.

phosphor. This emission from the phosphor is then imaged onto a detector. In the past a film pack was used, in conjunction with an image intensifier, but nowadays, most commonly, we have a CCD or other electronic camera.

Both the temporal and spatial resolution as well as the spectral response are affected by the photocathode material used, which must have a low work function to allow optical photons to remove photoelectrons from the surface. It also needs to be semi-transparent to allow light falling from the source side to be able to release the photoelectrons on the side facing the grid. To prevent space charge building up, they must also be electrically conducting. Very thin (~100 Å) metal photocathodes are semi-transparent to optical radiation and can be used but their efficiency is low and, typically, the photocathode consists of a mixture of alkalis with the incorporation of a thin metallic layer to prevent charge build-up. For instance, the S20 cathode whose response is shown in figure 5.1 is composed of a Na-K-Sb-Cs mixture [3], whilst the S1 cathode is Ag-O-Cs, with a bulk made from a complex structure of Ag particles in a Cs_2O matrix [4]. The S1 has the advantage of a low work function that allows operation into the near infra-red but suffers from a low quantum efficiency, compared to the S20.

The temporal resolution depends on several factors. Firstly, the photoelectrons ejected from the photocathode have a small spread in energy. This leads to a spread in the time of flight down the tube. There will also be space charge effects that, at higher signal levels, where there is more charge, also leads to temporal broadening during the flight down the tube and limits the dynamic range of the signals recorded [5]. With a fast sweep unit and a suitable slit, resolutions of around 10 ps or better are readily achieved. Likewise, for spatial resolution, the electro-optic focussing system, phosphor and cathode all contribute. Typically, if we define the resolution as the ability to image a bar chart with 50% contrast, the resolution achievable is around 10 line-pairs/mm.

5.2 Optical pyrometry measurements

If we refer back to chapter 1 and figure 1.1, we can see the shock Hugoniot curves for both H and Fe. It is clear that, when shocks of pressures in the megabar (100 GPa) range are used to compress a sample, temperatures of up to 10^4 K and higher, can be readily generated. Not only can this produce melting of the sample, but it leads to significant emission in the optical regime, either in a transparent sample or as the shock-wave exits the rear of an opaque sample. The appearance of this shock emission has been used in many experiments as a signature of the shock reaching the rear of a sample [6, 7]. By measuring the time of emission, relative to the shock drive, usually with an optical streak camera, the shock speed can be inferred. This can be done by comparing the appearance of emission from the rear of the target to a timing fiducial created with the shock driving beam, as for example in [8]. Alternatively, a stepped target [9] can be used as long as the drive intensity is uniform over a large enough area to allow comparison of emission times across the spatially separated steps. This stepped target technique has the advantage that it can allow the shock speed, in a specific material, to be directly measured without having to factor in the delay due, for example, to an ablator layer.

An example of streaked optical emission data [10] is shown in figure 5.2, where a shock has been driven into a Be/Au/Quartz layered target. The Be is a low density ablator, the Au is the pre-heat layer. The shock pressure in the quartz decays from ~300 GPa to ~80 GPa in this data. We can see the effects of shock reverberation from layers of different shock impedance and when the decaying shock is no longer strong enough to melt the quartz, at about 120 GPa, there is also a rise in the emission, as shock energy goes into heating rather than a melting phase change.

In addition to indicating the shock break-out time, the streaked optical emission has been used to estimate the shock temperature, where the technique is often called streaked optical pyrometry (SOP). There are two modes in which this has been done, both of which exploit the fact that, at the high density of a compressed sample, the emission is generally almost black-body in nature. In the first method, the spectral shape of the emission in the optical regime has been recorded and a black-body curve has been fitted to find a spectral temperature, T. Using Wien's displacement law, we can see that the peak emission would occur at wavelength given by;

$$\lambda_{\max}(\mu\text{m}) = \frac{2897}{T} \tag{5.1}$$

where T is in Kelvin. We can see that for the optical responses of a streak camera seen in figure 5.1, we would, ideally, like peak emission to be in the 400–600 nm range and thus $T \sim 5000$ K or higher. For lower temperatures, not only is the peak emission shifted to longer wavelengths in the IR regime where response falls off rapidly, but, for a quasi-black-body, we also expect a T^4 scaling of the emission intensity. For this reason, SOP has generally been used to identify shock temperatures of order 3000 K or higher, for example see [11]. As noted already, figure 5.1, shows us the typical spectral response for optical streak camera tubes fitted with either S20 or S1 photocathodes. The S1 cathode has a response that extends out into the near-IR. However, this comes at the expense of a sensitivity, in the optical regime, a couple of orders of magnitude lower than the multi-alkali S20. In both cases, the lower wavelength response of the system will be generally determined by the optical materials used. For glass (for instance BK-7), the lower wavelength transmission generally cuts off at around 350 nm, whereas MgF_2 optics can be used down to well below 200 nm, but would then require suitable optics in the whole experimental imaging system.

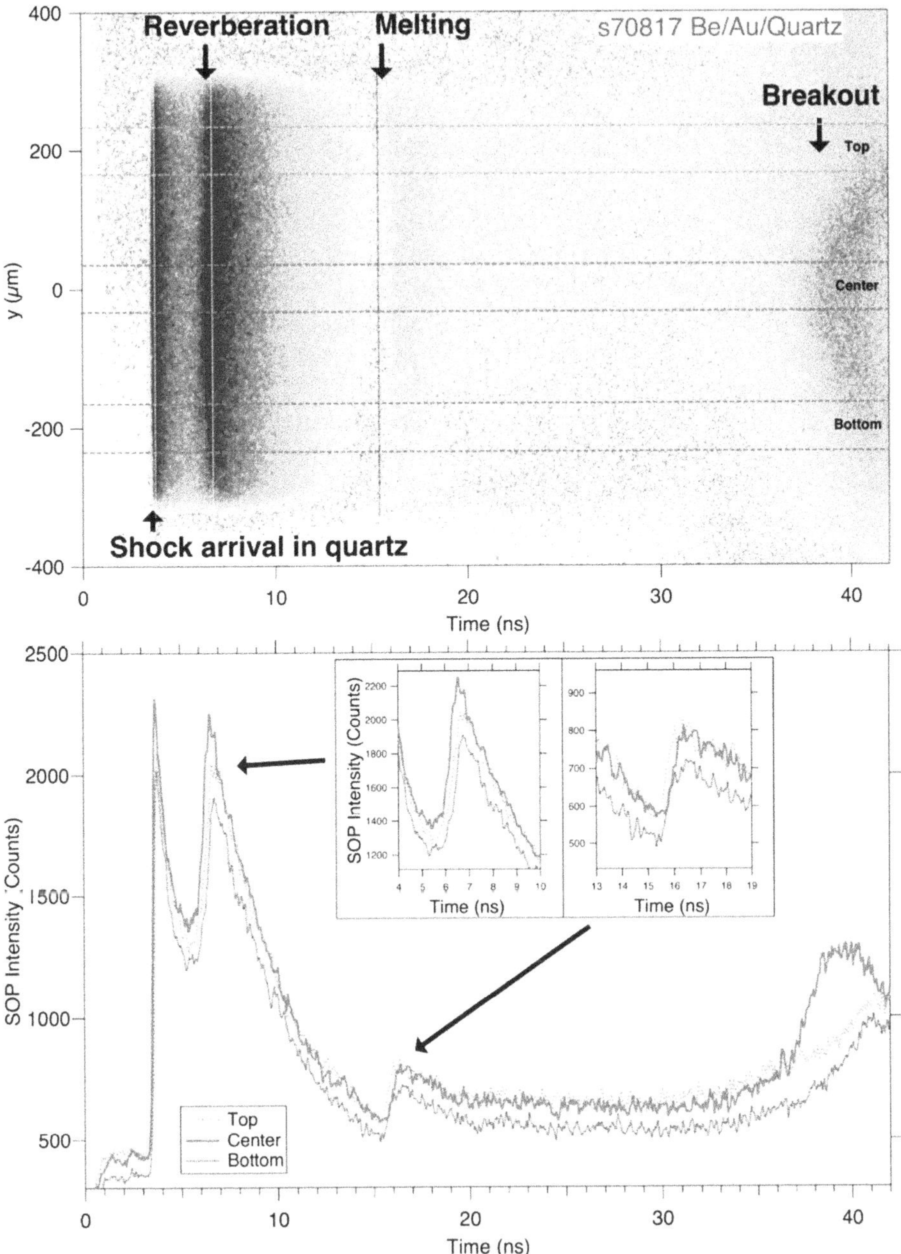

Figure 5.2. An example of streaked optical emission from the rear of a stishovite sample on a quartz substrate compressed with a decaying shock. The effects of shock reveberation and melting can be seen in the changes to the optical emission. Figure reproduced from [10], with permission of AIP Publishing ©2016 AIP.

In the absence of absolute calibration, we can obtain a spectral temperature where we match the shape of the spectrum to a black body at a given temperature, which is given by;

$$I(\lambda) = \frac{\varepsilon(\lambda)2hc^2}{\lambda^5}\frac{1}{e^{hc/\lambda k_B T} - 1} \tag{5.2}$$

where, T is again in Kelvin and $I(\lambda)$ gives the power radiated per unit area per steradian per unit wavelength. The emissivity, $\varepsilon(\lambda)$ is 1 at all wavelengths for a black body. In order to make a sensible fit it is, of course, necessary to have a clear measurement over a wide enough spectral window and we need to understand the relative response of the system used to record the data, including all the optics that bring the emitted light to the streak camera. This can be achieved by utilising a black body light source, typically a tungsten-halogen lamp [12], to make measurements and compare the raw output data with the expected spectrum for the source. Such sources can produce close to black-body output with spectral temperatures of 3000 K, although the emissivity is $\ll$1. The calibration can often be traceable back, for example, to the National Institute of Science and Technology (NIST) or other national laboratories responsible for maintaining such standards calibrations.

Sometimes, absolute calibration of the streak camera and optical imaging system is possible and the absolute emission at a single wavelength or multiple wavelengths is used to determine an effective emission temperature. As an example, a detailed discussion of the system implemented at the Omega laser facility is found in [13].

If the shock is driven in a transparent medium, then the emission can be measured before the shock has broken out. In these cases, care has to be taken to understand the optical transmission of the material ahead of the shock. An added complication is that the optical properties can be changed by pre-heating of the material ahead of the shock front. For non-transparent targets, such as metals, we only observe the optical emission from the shock when it exits the sample. At this point, as discussed in chapter 2, there is a rapid decompression of the target that can make interpretation of the emission temperature difficult and rapidly changing opacity effects will come into play. One way of addressing this issue is to use a transparent window on top of the sample. As the shock reaches the interface, into which the shock propagates, the emission is visible and the opaque sample is prevented from decompressing. However, solving the decompression issue introduces the problem of understanding the optical properties of the window material under pressure, and, as with a transparent sample, the issue of changes in optical properties with pre-heating. This is a significant issue as it is necessary for the window to remain transparent and for the transmission as a function of wavelength to be understood. In fact, the optical properties of window materials such as fused silica and quartz have been explored to pressures of 250–300 GPa and in the case of LiF, the refractive index has been measured in ramp compressions to 800 GPa [14]. It is not simply a question of the reaction of the window to pressure that is important, but the quality of the interface and thermal conduction into the window are also key parameters to be considered. For example, Huser *et al* [15] have studied the release of up to 600 GPa shocks from iron into a LiF window. In this work, they used a model of the thermal conductivity to understand the changes in the conditions within the LiF window as the shock traversed the interface.

It is important to bear in mind that the temperature we measure at the rear of a shocked sample is not the same as the shock temperature, T_s. As we have seen, an isentropic release state is created when the shock emerges into vacuum or a window. The temperature of this release state, T_r, is given by;

$$T_r = T_s \exp\left(-\int_{V_0}^{V} \frac{\gamma_G}{V} dV\right) \tag{5.3}$$

where, as before, γ_G is the, density dependent, Grüneisen parameter. As noted by Huser *et al*, for Fe release into LiF, there can be a 25% difference in shock and release temperatures.

5.2.1 Non-streak camera based pyrometry

The optical streak cameras described above are usually very expensive and yet temporal resolution down to the picosecond level may not be necessary. Other time resolved pyrometric techniques have been developed. For example, as part of the ion-beam WDM experiments, proposed in recent years, a diode based multi-channel system has been developed [16], in which the thermal emission is collected using a high collection efficiency spherical or parabolic mirror system and coupled to a fibre optic. The signal is transported into a spectral analyser where the fibre optic signal is split into six detector channels, each with an interference filter, in the 500–1500 nm range, to select a narrow wavelength range (around 20–40 nm). This range means that the peak black-body emission should be included as the temperature varies from 1500 to 6000 K. It is worth noting that the use of a focusing mirror arrangement to gather the emission avoids the use of transmissive optics and the chromatic aberrations, which would be difficult to deal with over such a wide spectral range. This leads to the ability to collect data over a relatively small spatial area, in the example shown in reference [16] better than 100 microns spatial resolution is expected.

The temporal resolution of this system is limited by the detector properties. For detection in the 500–900 nm range, Si based PIN diodes are suitable, whilst, for longer wavelengths, an InGaAs diode can be used. With a small area (1 mm^2) and a suitable bias (>10 V), rise times of 5 ns can be achieved. The time resolution could be made faster with higher reverse bias and can relatively easily be of order of a nanosecond, which is adequate for many cases, such as ion-beam heated WDM, which typically will have heating timescales in the order of 10s nanoseconds.

For some shock experiments, depending on pressure regime and material, the optical emission may be weak and it may prove difficult to get good data on the emission temperature and even an accurate time of shock break-out. In such cases, we can consider an active probing method, such as VISAR which does not measure temperature, but does give a signature of shock break-out and motion of the rear surface.

5.3 VISAR measurements

One of the most important shock diagnostics that is commonly run along with SOP is VISAR, which is an acronym for *Velocity Interferometry System for Any Reflector*. This is a technique, initially developed in the early 1970s [17, 18] that uses a probe laser to monitor the velocity of a surface, such as the rear of the target when a shock breaks out, and many major facilities will have it installed as a standard diagnostic. One of the key advantages of this system over earlier interferometry techniques is that it does not require a highly polished, mirror-like surface; hence the name. There are several different possible configurations, but details of a fairly typical design can be found in [19].

The way in which VISAR is usually implemented is that a telescopic lens arrangement is used to focus the probe laser, via a beamsplitter, onto the rear surface of the shocked sample, ideally in a uniform focal spot which is often larger than the expected lateral extent of the shock. The reflected light is collected by the same lens system and directed back through the beamsplitter, into an interferometer with a layout similar to that shown in figure 5.3(a).

The purpose of having an étalon in one arm is that this allows us to introduce a time delay between the arms, but at the same time allows the target surface to be imaged onto the output beamsplitter for both arms This is important since this allows good contrast for the fringes without relying on spatial coherence of the reflected light, thus allowing non-specularly reflecting target surfaces to be used. Initially both arms of the interferometer can be set-up, without an étalon, to have the same path-length and this can be checked by looking for white light fringes using a broad-band input source. The étalon can then be inserted, as in the figure, introducing the time delay in one arm which depends on the refractive index and thickness, h, of the étalon. The étalon also shifts the focus and, in order to re-establish the proper overlap of the output images, the mirror in this arm then needs to be moved a further distance such that the length of the delayed arm is larger by;

$$\Delta L = h\left(1 - \frac{1}{n_0}\right) \tag{5.4}$$

where the refractive index, n_0 is for the probe wavelength, λ_0. A detailed derivation and discussion of these requirements can be found in [20]. The combined time delay between the arms due to this shift and the refractive index is given by;

$$\tau = \frac{2h}{c}\left(n_0 - \frac{1}{n_0}\right) \tag{5.5}$$

With étalons typically between 2 and 60 mm thick, the time delay is of order 10–300 ps. For this reason, although the spatial coherence of the probe is not critical, temporal coherence is needed and should be longer than the longest likely time delay for the experiment. As discussed in [20], the time delay, between the two arms means that, for a moving reflector, the position of the surface at the time the probe is reflected, is different for the two arms, introducing a phase difference at the output.

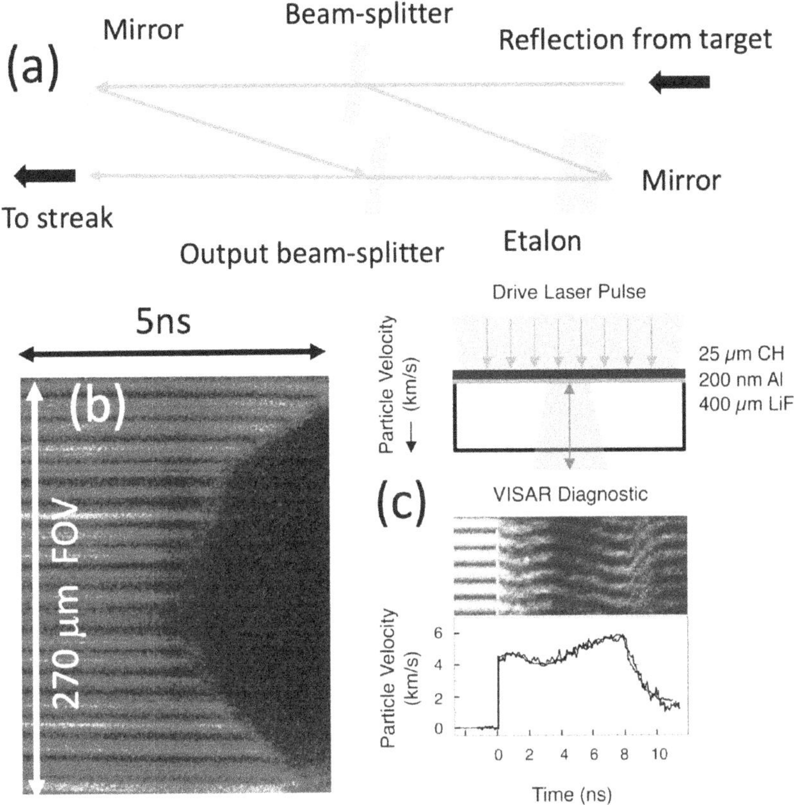

Figure 5.3. (a) Schematic of the interferometer at the heart of a VISAR system. (b) Example data showing break-out from a laser-shocked Al foil. The fringes disappear as the shock breaks out and the surface forms a highly absorbing plasma. (c) Data for a shock breaking out into a LiF window. We see the initial jump as the shock passes through the interface, followed by evolution of the velocity of the Al/LiF interface. Image reprinted from [24] with the permission of AIP Publishing ©AIP Publishing 2017.

For a steady velocity, this is constant, proportional to the velocity, and we don't see any fringe shift during the experiment. However, if the velocity changes during this delay time, the Doppler shifted wavelength for the two arms is different and this introduces a phase shift and thus movement of fringes. The size of the time delay sets a limit to the resolution with which changes in velocity can be determined. The Doppler shifts in question are generally small. For the VISAR case, where the source and detector are in the same laboratory frame, this is given by:

$$\frac{\lambda}{\lambda_0} = \frac{1 - v/c}{1 + v/c} \tag{5.6}$$

In general, we will be concerned with situations where $\frac{v}{c} \ll 1$ and we can simplify to

$$\frac{\lambda}{\lambda_0} \sim 1 - \frac{2v}{c} \tag{5.7}$$

Despite the small values of velocity compared to c, the device can be very sensitive. We can see this, with an example, where we assume some typical values of 5 mm for the étalon thickness and 1.46 for its refractive index at a probe wavelength of 532 nm. Equation (5.5) gives a difference of nearly 26 ps between the reflected light leaving the target surface for the two arms As we have noted, for a steady velocity, we see no fringe shift. However, if the velocity jumps, for example, from zero to a value typical for laser-driven shock physics, of 10 km s^{-1}, within this period, then the arm without an étalon sees the initial wavelength and the delayed arm sees a Doppler shifted wavelength. Since the fixed time delay corresponds to around 14 500 wavelengths, we can see that a change in wavelength of only 1 part in 14 500 is needed to change the phase at the output by 2π radians and thus see a whole fringe shift during the experiment. This, as it happens, is the change in wavelength provided by the assumed 10 km s^{-1} velocity. In general, the sensitivity is usually expressed as the velocity per fringe (VPF):

$$\mathrm{VPF} = \frac{\lambda_0}{2\tau(1 + \delta)} \tag{5.8}$$

where the time-dependent velocity, $v(t)$ can be extracted by multiplying this term by $N(t)$, the number of fringe shifts measured at a given time, t. The term δ is a small correction for the fact that the refractive index changes slightly with the Doppler shifted wavelength [18]. This term is typically $\ll$1 and makes a few percent difference to the deduced velocities. There are other small corrections that deal with the angular spread of the reflected light as it is collected by a fast lens [19, 20]. As we can see from equation (5.8), a thicker étalon is more sensitive, requiring a smaller velocity to induce one fringe shift. For a quartz étalon, the 2–60 mm range gives sensitivities ranging from about 0.9 to 25 km/s. If we set the sensitivity to be higher, in order to get more accuracy in the measured velocity, we run the risk of having shifts of more than one fringe. This would lead to ambiguity in the derived velocity and, for this reason, it is common to employ two systems with different fringe sensitivities. There will generally be only one value of fringe shift, and hence velocity, that is compatible with both arms and this removes the ambiguity.

The fringe shifts caused as the shock emerges from the rear of the sample can be recorded on an optical streak camera, for example by imaging the output beamsplitter onto the photocathode, giving spatial resolution across a line on the target surface as well as temporal resolution. Analysis of the fringe shift can be used to extract the velocity history of the surface. As noted above, for best fringe contrast, the focal plane of the images of the target from the two arms should be overlapped on the output beamsplitter. In operation, the target surface moves position. For a typical laser-drive shock experiment, we expect movement of order 10 μm during the experiment. Happily, this is generally less than the focal depth of the imaging system and does not prevent decent contrast. The effect of the surface movement has been considered in some detail by Sweatt [21]. Since an optical streak camera uses a slit to obtain time resolution, only a narrow line-out across the fringe image is likely to be recorded. In some cases, a cylindrical lens is used to collapse the fringe pattern onto the slit, enhancing the signal level.

In figure 5.3(b) we see an example of when a shock breaks out from the rear of a simple opaque target. The rear surface rapidly decompresses, forming a low-density plasma/vapour that strongly absorbs the probe and the fringes disappear. This is the basis of an active shock break-out (ASBO) diagnostic that indicates the time at which the shock reaches a rear surface, although this type of diagnostic can be implemented in a simpler manner, without the need for a VISAR interferometer [22], where the drop in reflectivity is an indication of shock break-out.

In obtaining the data for figure 5.3(b) a 100 μm focal spot was used in driving a shock through a 25 μm thick foil. The field of view is larger than the shock region and we see the effects of non-planar shock propagation in the curved break-out line. We can also see evidence of some motion in the fringes prior to their disappearance. This is a signature of pre-heating that can heat the solid ahead of shock break-out, causing motion of the rear surface even as it maintains a steep density gradient capable of reflecting the probe laser. Such data has been used to measure pre-heating in shock physics experiments [23], and this is an important diagnostic because, as we have discussed, use of the Rankine-Hugoniot equations to determine shock conditions is predicated on knowing the initial sample conditions prior to shock compression.

If the shocked sample we are using is initially transparent, then, at high enough pressure, the shock can turn the sample into a reflective, metallic state and the surface we measure the velocity of is the shock front. For an opaque target, we need to wait until the shock breaks out of the rear. As we discussed in chapter 2, the velocity of the surface is then governed by the isentropic release. As we have also noted, the expansion to vacuum creates a low density, highly absorbing vapour. This is where the use of a window on the rear of a target can be useful. As an example, the data in figure 5.3(c) comes from an experiment in which a shock is driven through a CH ablator layer which is coated in Al to form a reflective surface for the VISAR probe [24]. The shock breaks out into a LiF window and we see the initial sharp jump in the fringes as the shock reaches the Al layer. In this case, the shock did not cause the LiF to lose transparency and we see the evolution of the velocity of the Al/LiF interface, which is governed by the continuity of the particle velocity across the interface. Combined with use of well characterised stepped targets, to make accurate shock velocity measurements, the use of windowed targets can allow simultaneous determination of both u_s and u_p, closing the Rankine–Hugniot equations.

Some key issues arise in the use of transparent targets and windows. Firstly, the refractive index will alter the apparent velocity measured in the VISAR. This can be corrected for [19, 20], if the refractive index of the window is known. However, as we noted in the discussion of optical pyrometry above, the optical properties of the window will change on shock compression and may be affected by pre-heating before shock compression. In some cases, it may be that the window changes from a transparent dielectric into a reflecting metallic state as it is compressed, thus the probe will monitor the velocity of the shock in the window rather than the velocity of the interface between sample and window.

So far, our discussion has centred around implementing line-imaging VISAR using optical streak cameras, which give excellent temporal resolution combined with spatial resolution along one direction, but are expensive, especially if two channels are used. There are alternative implementations [25] that have been constructed with the probe delivered to target via a fibre optic coupled to a lens. In these designs, we do not gain a spatial profile but sample a small spatial region defined by focusing of the probe. The reflection of the probe is collected by the lens and fed into the fibre optic. Fibre splitters are used to create two channels with a relative delay provided by different fibre length. In this case we do not see a spatial image of an interference pattern as in figure 5.3, but a time trace of oscillations in the light intensity as two channels of the interferometer 'beat'. We gain the advantage of a cheaper system at the expense of losing the spatially resolved profile.

5.4 Frequency domain interferometry

In some experiments with WDM generation, it is possible that a sample with a very steep density profile is created. These can include experiments where a thin foil is volumetrically heated with x-rays or irradiated with an ultra-short-pulse laser pulse. If the scale-length, L, of the density is smaller than an optical probe wavelength, for example;

$$\left(\frac{1}{n_e}\frac{\partial n_e}{\partial z}\right)^{-1} = L < \lambda_{\text{probe}} \tag{5.9}$$

then we can expect good specular reflection from the critical density surface. In such cases, we can think about Fourier Domain Interferometry (FDI), which is an optical diagnostic technique [26–29] that can measure the speed of expansion of the critical density surface and this, in turn, can be compared to simulations to confirm the hydrodynamic behaviour.

This diagnostic depends on using a laser pulse with a broad spectral width, such as is found in a chirped pulse amplified (CPA) laser [30], to translate spectral resolution into temporal resolution. As with some other types of diagnostic, there are several variations on how FDI is implemented. An early implementation is illustrated in figure 5.4.

In this implementation [31] the WDM sample is created by heating with an ultra-short pulse laser (77 fs, 620 nm wavelength) focused to 3×10^{15} W cm^{-2} onto a solid Al target. A separate 40 fs pulse at 590 nm was synchronised to the pump to create a reference and a probe pulse. This is done by passing the second, separate, pulse through a Michelson type interferometer with unbalanced arms, which produced two pulses that are collinear but delayed, by typically a fraction of a picosecond to a few picoseconds. As we see from the figure, if the reference pulse is reflected from the surface before the pump pulse creates the sample, then the reflected wave-front only has a uniform phase shift that leaves the pulse, in all essentials, unchanged. The probe pulse, however, is affected by the changed conditions. Contributions to the phase include the usual phase jump that occurs on reflection from a surface, which is dependent on the polarisation of the probe. There is also a phase shift due to propagation of the probe in the under-dense plasma that is created on the surface,

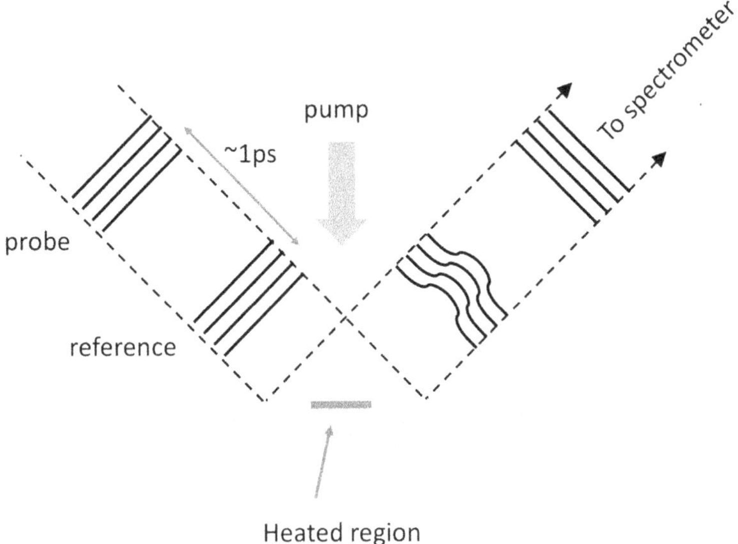

Figure 5.4. Schematic of the FDI technique, similar to that used by Geindre *et al* [31]. The focal spot of the pump beam is of order 10 microns, whilst the probe and reference pulses are over 40 microns in diameter.

as well as the Doppler shift induced by the motion of the critical density surface from which the probe pulse is reflected.

After reflection from the surface, the probe and reference pulses are imaged onto the input slit of a grating spectrometer. The spectral width of a 40 fs pulse is of order 30 nm and, within the spectrometer, the path difference between the frequency components means the two pulses are stretched to overlap temporally and can interfere to create a fringe pattern. Effectively, the spectrometer is Fourier transforming the pulses from the time domain to the frequency domain, hence the name of the technique. If the spectral dispersion of the spectrometer is given by $S = \partial x/\partial\omega$, then the spatial frequency of the fringes is given by $2\pi S/\Delta t$ where Δt is the delay between the pulses. The intensity pattern as a function of frequency in the measurement is given by;

$$I(\omega) = I_0(\omega)[1 + R_p + 2\sqrt{R_p}\cos(\omega\Delta t + \phi)] \tag{5.10}$$

where ϕ is the phase shift of the probe beam on reflection and R_p is the reflection coefficient, relative to the initial intensity of the probe pulse. Taking the Fourier transform of this pattern allows the extraction of the phase change and the reflection coefficient. In this implementation we are measuring the change in phase and reflectivity between the cold sample and the evolving plasma surface at the time of reflection for the probe pulse. This is sometimes called the absolute mode. However, we cannot make the time difference between the pulses arbitrarily large since this will increase the number of fringes and there are practical limits to spatial resolution. However, we can operate in a mode where both reference and probe pulse are incident after the pump pulse and both sample the expanding critical density surface.

This will monitor the relative phase changes between reference and probe resulting from evolution of the plasma over the small delay between them, but at a time long after the pump pulse. This is often called the relative mode.

In order to prevent heating of the target by the probe, the intensity on target of the probe is typically less than 10^{12} W cm^{-2}, which also prevents the ponderomotive force of the probe from affecting the expansion of the critical density surface. In order for this diagnostic to work, the surface optical quality of the target should be good and often considerable effort has gone into manufacturing the targets. For example, Shepherd *et al* [28] coated silicon nitride onto optical quality Si substrates before etching away the silicon and adding coatings of Al and C to create optically flat sandwich targets.

Typically, in implementations of FDI, the probe and reference beam are designed to illuminate, not just the heated region but an unheated region around it. This means that, in taking data, we have a set of unshifted fringes to give a reference for measured fringe shifts. As you may guess, this may restrict the use of this diagnostic to experiments in which it is practical to achieve this. For example, with a radiatively heated foil it may not be possible to restrict the heated sample to a small enough region to allow for good optical flatness of the target.

An alternative implementation of FDI has been used by Benuzzi-Mounaix *et al* [29] to explore the rear surface of the sample compressed by a laser-drive shock. In this implementation, the probe and reference pulse, are chirped pulses, as before, but used in an only partially compressed state with a pulse duration, in their case, of 75 ps. The pulses were focused onto the rear of a sample target with a pulse separation of only 10 ps. This ensured good temporal overlap and a null shot with no shock drive was used on the target surface to establish a clear interference pattern when the pulses were imaged into a high resolution optical spectrometer. The pump beam could then be timed so that the shock break-out from the rear of the sample occurred during illumination by reference and probe beams. As the rear surface evolved and expanded, there would be a phase change induced between reference and probe.

5.5 Reflectivity measurements

We noted, in chapter 1, that the dc electrical resistivity (and hence dc conductivity) is closely connected to the structure factor of WDM and thus measurements of such bulk properties are of interest in deducing the microscopic properties of WDM. There will also be an ac conductivity which can be connected to the optical properties of WDM and there are numerous papers dealing with plasma conductivity [32–34], for dense plasmas, where degeneracy and strong coupling can play a key role. The conductivity is closely connected to the plasma dielectric function and thus the refractive index. This means that experiments to measure reflectivity of a WDM sample with a laser of wavelength, λ can be useful tests of modelling.

In fact, amongst the early optical diagnostics applied to WDM research were relatively simple reflectivity measurements where a short pulse optical laser is reflected from a slab-like sample. Initially these were measurements of the self-reflection of the pulse used to heat a solid surface [35, 36]. If the sample has a sharp density gradient, L, as in equation (5.9), then we can use the Fresnel equations with measurements of the reflectivity for both S-polarised and P-polarised light to deduce a value for conductivity. In the limit of a sharp interface between the sample and vacuum, the reflectivity, at normal incidence, is given by;

$$R = \left| \frac{\sqrt{\varepsilon(\omega)} - 1}{\sqrt{\varepsilon(\omega)} + 1} \right|^2 \tag{5.11}$$

where the complex dielectric constant is related to the complex refractive index via $n(\omega) = \sqrt{\varepsilon(\omega)}$ and is connected to the electrical conductivity, $\sigma(\omega)$ by;

$$\varepsilon(\omega) = 1 + \frac{4\pi i \sigma(\omega)}{\omega} \tag{5.12}$$

For a Drude like, ac conductivity, without contributions from atoms but just free electrons, we have the Boltzmann–Drude equation;

$$\sigma(\omega) = \left(\frac{n_e e^2}{m_e} \right) \left\langle \frac{\tau}{1 - i\omega\tau} \right\rangle \tag{5.13}$$

where τ is an energy dependent electron–ion collision time, which will depend on the degeneracy and temperature of the sample [35] and, in the above, τ is thermally averaged over the electron distribution including the effect of degeneracy. In many, if not most, cases, we will have a scale-length generated as the sample expands. In these cases, we will propagate the laser in a gradient and usually we would seek a solution by modelling using a wave-solver;

$$\frac{d^2 E_x}{dx^2} + \frac{\omega^2}{c^2}\left(\varepsilon(\omega) - \sin^2\theta \right) E_x = 0 \tag{5.14}$$

for s-polarised light, where in this case, the electric field is perpendicular to the density gradient always. For p-polarised light we can use;

$$\frac{d^2 B_x}{dx^2} + \frac{\omega^2}{c^2}\left(\varepsilon(\omega) - \sin^2\theta \right) B_x - \frac{1}{\varepsilon(\omega)} \frac{d\varepsilon(\omega)}{dx} \frac{dB_x}{dx} = 0 \tag{5.15}$$

where it is the B-field that is perpendicular to the gradient and, in both cases, we have assumed the laser is incident in the $y - z$ plane at an angle θ to the target normal which is given by the direction, z. In principle, we should expect that there will be a gradient in the conditions in the z-direction and the numerical solution to these equations will give the reflectivity if we have a model of the conductivity spanning the relevant parameter space.

An important requirement of experiments of this type is that there should not be a significant under-dense plasma before the probe laser reaches the WDM region of the sample. The absorption of the probe in the under-dense plasma and the reflection from the plasma critical density surface would obscure the data of interest. It is for this reason that this type of diagnostic has mostly centred around ultra-fast heating and probing experiments.

In earlier experiments, such as Price *et al* [35] the laser pulse acted both to create the WDM sample and to be the probe whose reflectivity was measured. The pulses were ultrashort, of order 100 fs, limiting the time for motion of the surface and development of an under-dense plasma. Nevertheless, temporal averaging in such an experiment was needed to compare theory and experiment. This was done by using the LASNEX hydrodynamics code where the electron–ion collision time was calculated using the model of Lee and More [32]. Good agreement with experiment was found over a wide range of incident intensity. In later experiments, an isochorically heated thin foil has been used, as in [37]. This is typically Al or Au and only a few 100 Å thick. Irradiance with an ultra-short pulse (<100 fs) laser is used to heat the foil relatively uniformly. This is possible because the range of the heated electrons is of the order, or longer than, the foil thickness.

In experiments with optical radiation, the conductivity measured is an ac conductivity at the frequency of the laser. In order to fully understand the properties of WDM, we naturally wish to extend such measurements to lower frequencies, ideally down to the dc level. If we look at the simple Drude model of conductivity, the ac and dc conductivities are related by;

$$\sigma(\omega) = \frac{\sigma_{dc}}{1 - i\omega\tau} \tag{5.16}$$

For a metal we expect τ to have a value between 10 and 100 fs. For a typical short pulse Ti:Sapphire laser operating at 800 nm, this means $\omega\tau \gg 1$. To get to the dc conductivity we must use a much lower frequency. Towards that end, experiments more recently have been extended to the infra-red and even the THz regime [38]. The THz frequency range is typically defined as 10^{11}–10^{13} Hz (0.03–3 mm wavelength). There are several methods of generating THz radiation but a common method for intense short pulses is to use the phenomenon of optical rectification. In this technique, an intense short pulse (~100 fs) laser is focussed onto a suitable crystal [39], for example zinc telluride (ZnTe) which can be typically ~mm in thickness [40]. The electrons, in such a crystal, are in an asymmetric potential well and, as a result, when subject to the oscillating electric field of a laser, a quasi-dc polarisation is generated. Such ultra-short laser pulses have a broad band of frequencies and this leads to a beating of polarisation that leads to strong emission in the THz regime. The pulse duration is typically sub-ps and the spectral width of the pulse is typically inversely proportional to the temporal width of the laser pulse, and the peak THz frequency is dependent on the optical wavelength. For ZnTe, the best efficiency is found in the range 0.5–3 THz. This means that for the range of collision times given above, we can now access a regime where $\omega\tau < 1$.

Because WDM samples tend to be small, it is desirable to focus the THz radiation to a small spot. Happily, since THz spectroscopy is an important technique in the measurement of crystal properties, and as a non-ionising probe in chemistry and biology, there are suitable THz optics available. These tend to be metallic reflective optics because, although focussing with a plastic lens such as Teflon is possible, the latter tends to include a high degree of absorption and dispersion. The reflective optics are usually in the form of an off-axis parabolic reflector. Such optics can also

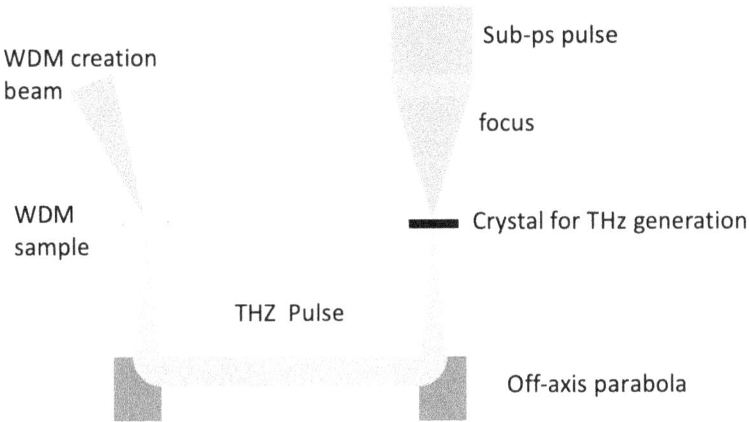

Figure 5.5. A generic experimental arrangement where a short pulse laser is split into two, one arm to drive a THz generation crystal and the other to drive a WDM sample. Electro-optical sampling of a transmitted or reflected THz pulse can then be used to analyse the properties of the sample, by observing the changes made to the THz pulse in amplitude and phase.

be used to collect radiation from the laser-crystal source and collimate it for transport before refocusing on the WDM sample with focal spot of order 1 mm, for example as seen schematically in figure 5.5.

Reflected or transmitted THz radiation can also be collected via such optics and detected with electro-optic sampling techniques [41, 42]. Typically, these techniques involve co-propagating a short pulse laser with the THz pulse through another electro-optical crystal which can, for example, be the same type used for the THz generation. The THz pulse induces birefringence via the Pockels effect and this causes rotation of the linear polarisation of the optical pulse. Analysis of the polarisation using a quarter wave plate and a Wollaston prism allows the intensity of the THz pulse to be determined as long as the rotation is kept less than 90 degrees. By using an echelle to split the optical pulse, derived as a pick-off from the main pulse driving the THz source and WDM sample, into many time-delayed parts, Ofori-Okai and colleagues [38] have been able to make single shot reflectivity measurements of a WDM sample.

References

[1] Bradley D J, Liddy B, Sibbett W and Sleat W E 1971 *Appl. Phys. Lett.* **20** 219–21

[2] McLean E A 1967 *Appl. Opt.* **6** 2120

[3] Townsend P D 2003 *Contemp. Phys.* **44** 17–34

[4] Gex F, Alexandre R, Horville D, Cavailler C, Fleurot N, Nail M, Mazataud D and Mazataud E 1983 *Rev. Sci. Instrum.* **54** 161–4

[5] Niu H and Sibbet W 1981 *Rev. Sci. Instrum.* **52** 1830–6

[6] Burnett H H, Josin G, Ahlborn B and Evans R 1981 *Appl. Phys. Lett.* **38** 226

[7] Cottet F, Romain J, Fabbro R and Faral B 1984 *Phys. Rev. Lett.* **52** 1884

[8] Cauble R, Phillion D W, Hoover T J, Holmes N C, Kilkenny J D and Lee R W 1993 *Phys. Rev. Lett.* **70** 2102

[9] Coe S E, Willi O, Afshar-Rad T and Rose S J 1988 *Appl. Phys. Lett.* **53** 2383

[10] Millot M 2016 *Phys. Plasmas* **23** 014503
[11] Hicks D G, Boehly T R, Eggert J H, Miller J E, Celliers P M and Collins G W 2006 *Phys. Rev. Lett.* **97** 025502
[12] Zhiyu H *et al* 2019 *High Power Laser Sci. Eng.* **7** e49
[13] Miller J E *et al* 2007 *Rev. Sci. Instrum.* **78** 034903
[14] Fratanduono D E *et al* 2011 *J. Appl. Phys.* **109** 123521
[15] Huser G, Koenig M, Benuzzi-Mounaix A, Henry E, Vinci T, Faral B, Tomasini M, Telaro B and Batani D 2005 *Phys. Plasmas* **12** 060701
[16] Ni P A *et al* 2008 *Laser Part. Beams* **26** 583–9
[17] Barker L M 1972 *Exp. Mech.* **12** 209–15
[18] Barker L M and Schuler K W 1974 *J. Appl. Phys.* **45** 4789
[19] Celliers P M, Celliers P M, Bradley D K, Collins G W, Hicks D G, Boehly T R and Armstrong W J 2004 *Rev. Sci. Instrum.* **75** 4916
[20] Dolan Daniel H 2006 Foundations of VISAR Analysis *Sandia Report* SAND2006-1950
[21] Sweatt William C 1992 *Rev. Sci. Instrum.* **63** 2945–9
[22] Ng A, Parfeniuk D, Celliers P, DaSilva L, More R M and Lee Y T 1986 *Phys. Rev. Lett.* **57** 1595–8
[23] Shu H, Fu S, Huang X, Wu J, Xie Z, Zhang F, Ye J, Jia G and Zhou H 2014 *Phys. Plasmas* **21** 082708
[24] Brennan Brown S *et al* 2017 *Rev. Sci. Instrum.* **88** 105113
[25] Levind L, Tzach D and Shamir J 1996 *Rev. Sci. Instrum.* **67** 1434–7
[26] Tokunaga E, Terasaki A and Kobayashi T 1992 *Opt. Lett.* **17** 1131–3
[27] Blanc P, Audebert P, Falliés F, Geindre J P, Gauthier J C, Dos Santos A, Mysyrowicz A and Antonetti A 1996 *J. Opt. Soc. Am.* B **13** 118–24
[28] Shepherd R *et al* 2001 *J. Quant. Spectrosc. Radiat. Transfer* **71** 711–9
[29] Benuzzi-Mounaix A, Koenig M, Boudenne J M, Hall T A, Batani D, Scianitti F, Masini A and Di Santo D 1999 *Phys. Rev.* E **60** R2488–91
[30] Strickland D and Mourou G 1985 *Opt. Commun.* **56** 3205–8
[31] Geindre J P 1994 *Opt. Lett.* **19** 1997–9
[32] Lee Y T and More R M 1984 *Phys. Fluids* **27** 1273–86
[33] Dharma-Wardana M W C 2006 *Phys. Rev.* E **73** 036401
[34] Wierling A, Millat T, Redmer R, Reinholz H and Röpke G 2001 *Contrib. Plasma Phys.* **41** 263–6
[35] Price D F, More R M, Walling R S, Guethlein G, Shepherd R L, Stewart R E and White W E 1985 *Phys. Rev. Lett.* **75** 252–5
[36] More R, Yoneda H and Morikami H 2006 *J. Quant. Spectrosc. Radiat. Transfer* **99** 409–24
[37] Forsman A, Ng A, Chiu G and More R M 1998 *Phys. Rev.* E **58** R1248
[38] Ofori-Okai B K, Descamps A, Lu J, Seipp L E, Weinmann A, Glenzer S H and Chen Z 2018 *Rev. Sci. Instrum.* **89** 10D109
[39] Hangyo M, Tani M and Nagashima T 2005 *J. Infrared. Millim. Terahertz. Waves* **26** 1661–90
[40] Venkatesh M, Rao K S and Chaudhary A K 2015 *AIP Conf. Proc.* **1670** 020005
[41] Wu Q and Zhang X C 1995 *App. Phys. Lett.* **67** 3523–5
[42] Ibrahim A, Férachou D, Sharma G, Singh K, Kirouac-Turmel M and Ozaki T 2016 *Sci. Rep.* **6** 23107

IOP Publishing

Warm Dense Matter
Laboratory generation and diagnosis
David Riley

Chapter 6

Facilities for warm dense matter research

6.1 Introduction

In this brief chapter, we will present an outline of some of the important facilities used in warm dense matter (WDM) research. Due in part to the sample size considerations outlined in chapter 1, these are often large scale facilities that are found in national laboratories and creating them is a major undertaking, often involving dozens if not hundreds of scientists as well as engineers, other staff and, of course, large amounts of money. As we shall see, there is often a desire to combine facilities, for example large pulsed lasers are being combined with x-ray free electron lasers and ion-beam facilities to create flexibility in the experiments that can be performed.

6.2 Laser facilities

The methods we have discussed in this book include using laser facilities as a source of x-rays or as a driver of strong shocks. As was noted, we often need large facilities with of order 100 J or more per pulse delivered. For nanosecond duration pulses, the lasing medium of choice is often Nd:glass. For phosphate glasses, the laser wavelength is around 1.054 μm and in most large facilities the laser is either frequency doubled to 527 nm or tripled to 351 nm [1, 2]. This can be done with efficiency exceeding 50% and confers significant advantages in absorption of the laser light [3] and also the scaling of laser-plasma instabilities, allowing better pressure generation and x-ray generation for a given irradiance. Quadrupling the laser frequency has also been utilised but this makes it necessary to use focussing optics capable of transmitting in the UV, such as high-quality fused silica or quartz. The build up of damage, due to UV photons, is an issue to be contended with and it is not usual to use fourth harmonic for the main laser drive. The details of how a laser works can of course be seen in many text books, for example [4, 5] give excellent introductions. Here, we will concentrate on giving the new researcher an overall view, of typical laser facilities, suited to someone who uses a large facility but is not the laser operator.

Long pulses
For a great deal of WDM research, such as shock compression and x-ray volumetric heating, we are dealing with so called long pulse lasers of typically nanosecond duration. In the past, these have been generated from Q-switched oscillators seeded with a single longitudinal mode, for example [6]. The pulses generated could be of order 10–50 ns and then cut down to sub-ns duration with very high contrast ($\sim 10^8$) by use of Pockels cells. The shape of the pulse is determined by the rise time of the isolators and the depletion of the high energy amplifiers, and the rise time of the pulse would typically be shorter than the fall-off time.

More recently, newer technologies have been developed to give control over the delivered pulse shape. This can be important, for example, in designing pulses to drive a sustained shock, or a ramped compression to reduce heating and explore off-Hugoniot states. For such shaped pulses, the start of the laser system, or *front end*, can be a fibre modulated arbitrary wave-form generator that creates a temporally shaped pulse that is fed into the amplifier stages. For example, at the National Ignition Facility (NIF) [7], a diode laser is used to pump a fibre laser generating a CW beam at the desired wavelength. This is chopped into 100 ns pulses using an acousto-optic beam chopper and, after further fibre amplification, is further chopped down to a 45 ns square optical pulse. This is then passed through a lithium niobate electro-optic modulator. The level of light passed through this modulator varies in response to an electrical input. By creating a series of 140 separate electrical Gaussian shaped impulses of 300 ps width and spaced 250 ps apart, an essentially arbitrary waveform can be generated. A similar system is used on the VULCAN laser at the Rutherford-Appleton Laboratory, Central Laser Facility [8] where 300 pulses each of 200 ps width and separated in time by 100 ps each are used. The shape of the arbitrary pulse is not exactly the final shape desired as account is taken, for example, of the temporal effect of the amplifiers being depleted in the lasing process. In the case of NIF there are 48 separately defined pulses, allowing each quad of 4 beams to have its own unique pulse shape if desired.

The shaped seed pulses, typically at around nJ energy level, are then expanded and fed into an amplifier chain. Usually, the first stages are rod amplifiers ranging typically from about 9 mm diameter. After amplification, the beam is further expanded before being fed into larger rods in stages which can be up to around 60 mm diameter. This allows high energy without exceeding the damage threshold of the amplifier glass. However, there is a practical limit to the diameter of the rods that can used, for a couple of reasons. Firstly, the flashlamp or diode pumping that creates the necessary population inversion needs to reach the centre of the rod and, due to absorption, this starts to be difficult to achieve uniformly for rods larger than the roughly 60 mm maximum used. Secondly, especially if flashlamp pumping is used, the rods absorb a great deal of heat and dissipating this is important before a new shot is fired. This is due to the fact that the refractive index of the glass is temperature dependent and a non-uniform temperature profile can effectively act as a lens, self-focusing the beam and damaging the rods. For a rod amplifier this timescale gets longer the larger the rod and the time between shots can start to reach half an hour for the largest rods. An important way around this is the use of disc amplifiers. These are discs of amplifying media, in the case of the VULCAN laser up to 20 cm in diameter and angled at the Brewster angle. The discs are a couple of cm thick and so cooling is readily achieved with air blowers. Use of such amplifiers allows energies of a couple of hundred joules to be generated for pulse lengths of about a nanosecond. The shot rate, which is partly governed by the need to cool the amplifying medium, can be around 20 minutes in the case of kJ level facilities but can be longer for larger facilities, with typically 1–2 shots per day for the NIF.

In recent years, diode pumping has been developed, in which optical diode emission can be generated that more closely matches the spectrum for pumping the laser medium. This means the pumping is more efficient, with less accompanying heating, and much faster shot rates are possible. The DiPOLE diode pumped laser [9] to be coupled to the XFEL facility in Hamburg is projected to generate nanosecond duration, 100 J pulses in second harmonic at 10 Hz. This is certainly enough energy for driving strong shocks over a relatively large focal spot.

Short pulses

If we are doing experiments on fast electron or proton heating or the generation of a broadband betatron source, we will typically use a chirped-pulse amplified (CPA) laser. The higher energy amplifier stages are similar to longer pulse lasers and the difference is in the front-end and the addition of a pair of pulse compression gratings after the amplifiers. A CPA laser system typically would start with a mode-locked oscillator that creates a sub-picosecond train of pulses at a repetition rate typically in the 10s MHz. These might have ~100 fs duration and nJ level energy. Amplification of such pulses to high energy would quickly run into the problem of non-linear focussing of the beam in the lasing material. Instead, the technique of stretching the pulse by utilising the narrow but finite spread in wavelength of the pulses was developed by Strickland and Mourou [10]. Today, this is done with a pair of anti-parallel gratings that cause the path length of the pulse to depend, nearly linearly, on the wavelength, with longer wavelengths coming earlier in the pulse (called positive chirp). This means that coming out of the stretcher, the pulse can be stretched to a few hundred picoseconds. The pulse can now be amplified to higher energy with over 100 J per pulse readily achieved in several laser facilities [11]. After amplification, a pair of parallel gratings can now be used to recompress the pulse by providing a negative chirp. The gratings are used at low angles of incidence which means the fluence is below the damage threshold. However, this also means they have to be large, high optical quality, gratings and thus expensive. At NIF, the advanced radiographic capability (ARC) makes use of two quads of the 192 beam-lines to deliver CPA pulses of 1–50 ps duration with up to ~10 kJ [12, 13] in total. Short pulse delivery of kJ energies is proposed in the Laser Megajoule (LMJ/Petal) Facility under development at the time of writing [14]. In table 6.1 we list some parameters for a small selection of large laser facilities that are generally accessible worldwide. Some of these are undergoing upgrades and the parameters may change, but the table should give a flavour of what is available. As we can see, the

Table 6.1. Some approximate parameters, of the principal beams, in a selection of major laser facilities with some reference links that are not exhaustive. The long pulse facilities (LP) have often been upgraded over the years and, for example, all have programmable shaped pulses.

Facility	LP beams	LP pulses	Energy	SP beams	SP energy
Vulcan [8]	6	0.5–5 ns	1 kJ (2ω)	2 × 1 ps	100 J (1ω)
Omega [15]	60	0.5–20 ns	40 kJ (3ω)	2 × 1 ps	1 kJ
NIF [16, 17]	192	0.5–20 ns	1.8 MJ (3ω)	8 × 1–50 ps	10 kJ
LULI 2000 [18]	2	~1–15 ns	1 kJ (1ω)	1 × 1 ps	100 J
PHELIX [19]	2	0.7–10 ns	1 kJ (1ω)	1 × 0.5–20 ps	200 J
Gekko XII [20]	12	1–10 ns	6 kJ (2ω)	1 × 1.5 ps	2 kJ
Shenguan II [21]	8	~1 ns	3 kJ (3ω)	1 × 1.5 ps	2 kJ

provision of both long and short pulses in the same facility is common and allows for pump-probe type experiments.

6.3 X-ray free electron laser facilities

The possibility of free electron lasers (FEL) was considered in the 1960s and, in the decades following, they were realised experimentally (see Feldhaus *et al* for references [22]). In the 21st century, facilities operating in the XUV and x-ray regimes have come online [23, 24] with the LCLS facility providing lasing into the hard x-ray regime (1.5 Å) since 2009. This facility has made significant contributions to WDM science, particularly in experiments where x-ray scattering has been fielded as a diagnostic and the narrow collimation, short pulse duration, and bandwidth of the beam are huge advantages.

The basic mechanism for operation is that a highly relativistic electron beam is steered through an alternating magnetic field created by an undulator. This consists of two parallel linear arrays of magnets of alternating poles, as shown in figure 6.1. As the electrons oscillate in the magnetic field, they emit synchrotron radiation. For a long undulator, as used in an FEL, the interaction of the electromagnetic radiation with the electron beam causes bunching of the electrons. In fact, the undulator at LCLS consists of 33 separate sections of 3.4 m in length each, and the undulator hall is 170 m long.

This bunching means that there is a higher degree of coherence in the emission, resulting in a peak brightness of typically more than 10^{32} photons/sec/mm^2/mr^2 in 0.1% bandwidth. This is around 10 orders of magnitude higher than the peak brightness of a synchrotron. The units of brightness make it clear that what we are talking about is not just high energy in a short pulse, but a very narrow highly collimated beam with a narrow spectral spread. This is the self-amplified spontaneous emission (SASE) mode. A self-seeded pulse can be generated by passing the beam through a Bragg crystal, which is bypassed by the electron beam, using magnets to create a chicane. The x-ray pulse spectrum is narrowed by diffraction and is re-injected into the electron bunch after the chicane and seeds the lasing process, creating a bandwidth in the 0.01% regime.

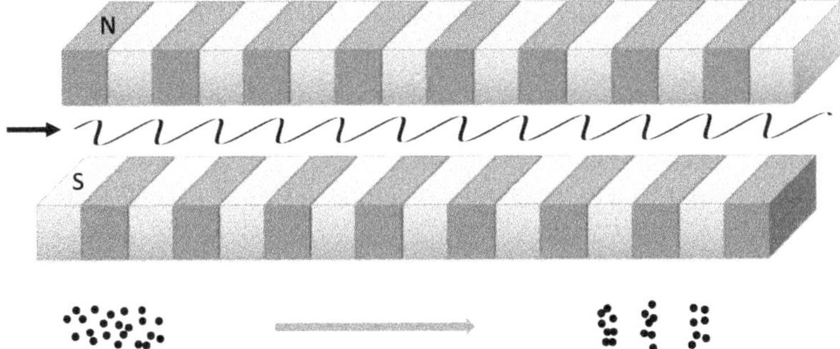

Figure 6.1. Schematic of an undulator operation. The electrons oscillate in the alternating magnetic fields and emit radiation. Initially the electrons are not bunched but emit incoherent radiation at the resonant wavelength of the undulator. As they interact with the radiation field they bunch together and start to emit coherently.

The wavelength of the free electron laser is determined by the optimum wavelength for energy exchange between the electron bunch and the electromagnetic field. For electrons of energy, E_e, this is given by;

$$\lambda = \frac{\lambda_u}{2n\gamma^2}\left(1 + \frac{K^2}{2}\right) \tag{6.1}$$

where

$$\gamma = \frac{E_e}{m_0 c^2} \tag{6.2}$$

and λ_u is the undulator period, and K is the dimensionless undulator parameter that is proportional to both the magnetic field and undulator period, with typical values varying between 1 and 5. Odd harmonics are also emitted corresponding to $n = 1, 3, 5$, etc. The harmonics are much weaker, with the third harmonic being about 1% of the fundamental in intensity, but both third and fifth have been observed, for example, in the FLASH XUV-FEL [24]. At the time of writing, the LCLS x-ray free-electron laser facility in Stanford, is supplied with an electron beam accelerated by a ~1 km section of Stanford LINAC linear accelerator to 16.5 GeV, and so $\gamma \sim 3 \times 10^4$. With undulator periods of only a few cm, we can see that resonant wavelengths in the Ångstom range are possible.

The bunching means that the pulses are of order 100 fs in duration. Both soft x-ray and a hard x-ray undulators are available operating between 0.2–5 keV and 1–25 keV respectively. In SASE mode the bandwidth varies between 0.1% and 0.2%, depending on energy and with $>10^{12}$ photons per bunch in a beam of order 10–20 µm across. The low divergence in the μm-radian range leads to a peak brightness of order 10^{31-34} ph/sec/mm^2/sr per 0.1% bandwidth, again, depending on energy. The low divergence aids in the ability to focus the FEL down to 0.1 µm, and it is possible to achieve intensities of 10^{20} W cm^{-2}. The XFEL facility in Hamburg [25] makes use of a 1.7 km linear accelerator to provide 17.5 GeV electrons. This can provide pulse parameters broadly similar to those of LCLS but with the use of superconducting magnets, the repetition rate can be 30 kHz compared to 120 Hz currently for LCLS. The LCLS II upgrade project underway at the time of writing, promises to upgrade the repetition rate to 1 MHz with a new 4 GeV superconducting linear accelerator. Of course, as with optical laser facilities, the provision of high repetition rates can pose challenges for WDM experiments. The targets are generally destroyed in the experiments and so if we are to use the higher repetition rates available, target replacement and debris issues need to be tackled. The high cost of some of the more complex targets means that the main benefit, in some cases, may be the much shorter duration of experiments rather than the larger datasets possible.

For both the XFEL and LCLS facilities, the provision of powerful optical lasers, synchronised to the x-ray beam is a key part of the capability. Nanosecond lasers with 10s of Joules pulse energies allow strong shocks to be created that can be probed by the x-ray beam [26] and short pulse beams have been used to create XUV high harmonics sources that have been used to probe x-ray laser heated foils at WDM conditions [27].

There are other x-ray free electron lasers operational, such as the SACLA facility in Hyogo Japan [28] operational since 2012, the SwissFEL [29] at the Paul Scherrer Institute, Switzerland and the Pohang Accelerator Laboratory (PAL) in Korea [30]. The scientific advances that have thus far been proven possible with these coherent, intense, short pulse x-ray sources, mean that it is likely that other future facilities will be developed.

6.4 Ion beam facilities

As noted in chapters 2 and 3, ion beams can be used to generate WDM via both volumetric heating and shock compression. They have been seriously considered as a source for inertial confinement fusion [31]. To do this, intense beams of ions with relatively short duration (<100 ns) need to be generated and focusable to a spot less than ~1 mm across.

Just as an oscillator forms the start of a large laser facility, the start of a large ion-beam facility is the ion source. In figure 6.2 we see a simple schematic of an ion beam

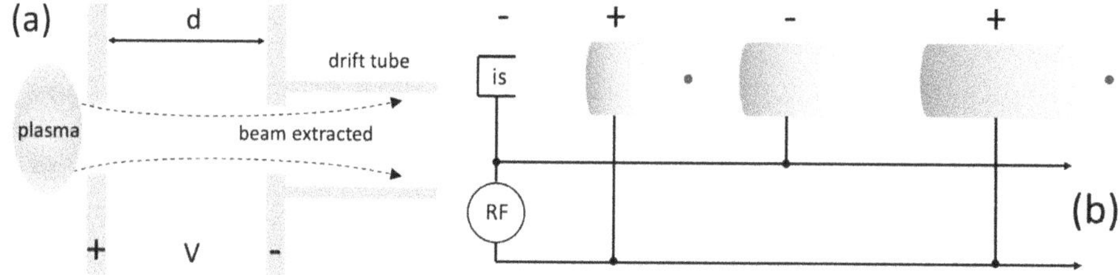

Figure 6.2. (a) Schematic of an ion source. The plasma can be generated in one of several ways, including electron cyclotron resonance, laser plasmas or Penning ionisation. This is the first stage of a large ion beam facility. (b) Simple schematic of the radio-frequency acceleration. The alternating potential, of the sequences of electrodes, pushes particles into bunches that are optimally timed to be pushed away from one electrode as they are pulled to the next.

source, which starts with a plasma that can be created by more than one method. For example, a vacuum arc or a laser plasma can be used and much research has been carried out into optimising these in terms of ion number and charge state distribution. Space charge limits the possible current density that can be extracted according to the Child–Langmuir limit;

$$I_{\max} = \frac{4\varepsilon_0}{9}\sqrt{\frac{2e}{m_e}}\frac{V^{3/2}}{d^2} \tag{6.3}$$

As an example, in the heavy-ion facility at GSI Darmstadt, the electron cyclotron resonance (ECR) plasma source has a current limit of approximately 1 mA. As described by Sharkov *et al* [32] multiple ion sources can be used to generate a high total ion-current. After the ion source, a radio-frequency quadrupole system can be used to select the appropriate ion species. This stage serves several functions. Firstly, it converts the essentially continuous ion flux, from the ion source, into bunches at the same energy, pulsed at the chosen radio frequency, usually 10s MHz. The ions are also accelerated to around MV voltages and this helps to reduce the space charge effects that can lead to defocusing of the beam.

The next stage is typically a linear accelerator that creates a higher energy beam. As an example of a facility that is capable of using ion beams for WDM research, the universal Linear Accelerator (UNILAC) at GSI, which is part of the Facility for Antiproton and Ion Research (FAIR) [33] has a capability to use several types of ion source to accelerate effectively any type of ion from protons to highly stripped uranium. After the ion source, the radio-frequency power source accelerates ions to about 120 keV/u (u = 1 atomic mass unit). Figure 6.2 shows in a simplistic way, how this happens. The accelerator consists of a series of cylindrical electrodes which are coupled, with the ion source, to the radio-frequency source, alternating 180° out of phase with both the previous and the next electrode. The acceleration is timed such that, as the voltage is positive on the ion source, it is negative on the first electrode, thus pulling ions forwards. For this to work, the ions have an optimum velocity to be synchronised with the voltage. For ions that are ahead, they will be slowed by the positive voltage still applied to the next electrode and thus be pushed back to the main bunch. The ions thus come in bunches at the radio frequency. At early stages, acceleration through a gaseous medium causes electrons to be further stripped from the ions thus increasing the ion stage and the maximum possible energy. These facilities can be very long, in the case of the UNILAC, it is around 120 m and the ions reach around 0.16c and about 11.4 MeV/u energy. Along the length of the linear accelerator, magnetic quadrupole lenses can be used to maintain the ions in the centre of the accelerator tube. At GSI the next stage is the SIS-18 ring accelerator which is 216 m in circumference and accelerates the ions whilst using magnets to keep them on a circular path. At each circulation, the ions are boosted by 80 kV. Using SIS-18 ion beams of up to 10^{10} U^{28+} ions at 300 MeV/u can be created [34] in pulses that are 50–100 ns in duration.

There are other heavy ion beam facilities where WDM research is being pursued. For example, stage II of the High Intensity heavy-ion Accelerator Facility (HIAF)

[35] in China, is being developed to deliver >10^{11} $^{238}U^{+34}$ ions in a 50 ns pulse at 0.84 GeV/u. At the time of writing, the Neutralized Drift Compression Experiment -II (NDCX-II) facility [36] at the Lawrence Berkeley National Laboratory, is being refurbished to allow delivery of a range of ion species at energies up to 1.2 MeV in nanosecond pulse durations. Focusability to less than 1 mm spots will allow the creation of WDM samples with ~1 eV temperatures.

Thus far, large scale ion-beam facilities have not featured as strongly in WDM research as large laser facilities. However, the advent of beams with large numbers of highly energetic ions has led to a large number of ideas and proposed collaborations such as the *High Energy Density Matter Generated by Heavy Ion Beams* (HedgeHob) collaboration [37] that, at the time of writing, includes participants from over 40 institutions in over a dozen countries. As discussed in chapter 3, such facilities may prove to have some advantages over laser facilities, such as the ability of ions to volumetrically heat higher Z samples without the need for a hot highly emitting plasma present in the experiment. The longer pulse durations, typical of ion beams, may provide experiments driven closer to equilibrium conditions, provided, of course, that large enough samples can be used that hydrodynamic expansion timescales can be suitably long. As with the x-ray free electron facilities, the addition of powerful synchronised optical lasers with the ions beams will add greatly to the versatility of the facility and at the time of writing the first experiments with the PHELIX laser [19] synchronised to upgraded ion beams capable of generating WDM relevant conditions are being planned.

6.5 *Z*-pinch facilities

In chapter 2, we discussed the generation of intense x-rays from Z-pinch facilities as a means to drive shocks. This type of device has a long history [38] and has in the past been considered as a potential fusion device. In recent years, many important WDM experiments have been carried out at Z-pinch facilities. Major Z-pinch facilities include the Z-facility at Sandia National Laboratories in New Mexico, the Angara 5-1 facility (see, for example [39] and citations within), in Russia and the Magpie facility at Imperial College, London [40]. A review of the physics can be found in papers by Ryutov *et al* [41] and Haines [38], but the core idea is that if a high current is passed through a cylindrical conductor a strong B-field is generated;

$$B = \frac{\mu_0 I}{2\pi r} \tag{6.4}$$

The fast rising current (timescales of order 100 ns) is initially confined to the surface by the skin effect. If we imagine a current of 1 MA passing and a fibre of 1 mm diameter, then the B-field at the radius of the wire is 400 T and the magnetic pressure is 64 GPa. At the same time, we expect significant resistive heating of the fibre, to the point that it can vaporise and form a plasma. If the fibre is narrower at some point along its length, then, as the current passing is continuous, the B-field is higher at the smaller radius and the fibre expansion is impeded more by the magnetic pressure. On the other hand, the current density, and thus resistive heating, is higher at the narrow point, and there is increased thermal pressure to balance the magnetic pressure. As we can imagine, this is potential a very unstable arrangement. and can lead to the so called *sausage instability* [38], which can form and break up the fibre. X-ray radiation generated by the hot dense plasma will tend to cool the fibre.

In x-ray generation experiments, wire arrays are used, where the current runs in parallel through up to several hundred wires, typically of order 10 μm diameter and 1–2 cm long. For the largest facilities, such as *Z*, the total current can be 20 MA and so each wire carries of order 100 kA. The wires are placed in a circular array and there may be two arrays nested within each other. The effect of the strong B-field generated in the parallel wires pulls them inwards where they can collide on axis. The dynamics are complex and subject to hydrodynamic instabilities and a pre-pulse current is used to condition the wires into a stable condition, before the main current pulse. The heating of the wires causes them to emit broadband x-rays with a quasi-Planckian spectrum [42]. The collision at the centre generates the main, few-nanosecond duration, intense pulse of x-ray emission, which can be extremely efficient, allowing up to 15% conversion of electrical energy to x-rays, with greater than 200 TW of x-ray power [38, 43].

6.6 Summary

As we have seen throughout this book, there are a wide range of techniques for the creation of WDM samples, and we have concentrated on presenting some of the most commonly used. Likewise, the facilities that can be used for WDM creation are diverse. They do all share the common goal of delivering a sufficient energy density, in a fast enough timescale, that it can create a sample at a high enough density and temperature to display WDM characteristics, and in a manner that is accessible to diagnosis. It is to be hoped that the material presented in previous chapters will still be useful to those who work on new types of facilities, that may be developed in the future.

References

[1] Craxton R S, Jacobs S D, Rizzo J E and Boni R 1981 *IEEE J. Quantum Electron* **QE-17** 1782–6

[2] Seka W, Jacobs S D, Rizzo J E, Boni R and Craxton R S 1980 *Optics. Comm.* **34** 469–73

[3] Garban-Labaune C, Fabre E, Max C E, Fabbro R, Amiranoff F, Virmont J, Weinfeld M and Michard A 1982 *Phys. Rev. Lett.* **48** 1018–21

[4] Davis Christopher C 2014 *Laser and Electro-Optics* 2nd edn (Cambridge: Cambridge University Press)

[5] Wilson J and Hawkes J F B 1993 *Optoelectronics: An Introduction* 2nd edn (Englewood Cliffs, NJ: Prentice-Hall)

[6] Ross I N *et al* 1981 *IEEE J. Quantum Electron* **QE-17** 1653–61

[7] Brunton G, Erbert G, Browning D and Tse E 2012 *Fusion Eng. Des.* **87** 1940–4

[8] Shaikh W, Musgrave I O, Bhamra A S and Hernandez-Gomez C 2006 *Central Laser Facility Annual Report for 2005-06* **19** 200–1

[9] Mason P *et al* 2018 *High Power Laser Sci. Eng.* **6** e65

[10] Strickland D and Mourou G 1985 *Opt. Commun.* **56** 192

[11] Danson C N, Hillier D, Hopps N and Neely D 2015 *High Power Laser Sci. Eng.* **3** e3
[12] Crane J K *et al* 2010 *J. Phys. Conf. Series* **244** 032003
[13] Tommasini R *et al* 2017 *Phys. Plasmas* **24** 053104
[14] Casner A *et al* 2015 *High Energy Density Phys.* **17** 2–11
[15] Boehly T R *et al* 1995 *Rev. Sci. Instrum.* **66** 508–10
[16] Hunt J T, Manes K R, Murray J R, Renard P A, Sawicki R, Trenholme J B and Williams W 1994 *Fusion Tech.* **26** 508–10
[17] Moses E I and Wuest C R 2005 *Fusion Sci. Technol.* **47** 314–22
[18] Benuzzi-Mounaix A *et al* 2006 *J. Phys. IV* **133** 1065–70
[19] Bagnoud V *et al* 2010 *Appl. Phys.* B **100** 137–50
[20] Kitagawa Y *et al* 2004 *IEEE J. Quantum Electron* **40** 281
[21] He X T *et al* 2016 *J. Phys. Conf. Ser.* **688** 012029
[22] Feldhaus J, Arthur J and Hastings J B 2005 *J. Phys.* B **38** S799–819
[23] McNeill Brian W J and Thomson Neil R 2010 *Nat. Photon.* **4** 814–21
[24] Ackermann W *et al* 2007 *Nature Photon.* **1** 336–42
[25] Altarelli M *et al* 2007 The European x-ray free-electron laser *Technical Design Report* DESY 2006-097
[26] Brown S B *et al* 2017 *Rev. Sci. Instrum.* **88** 105113
[27] Williams Gareth O *et al* 2018 *Phys. Rev.* A **97** 023414
[28] Kato M *et al* 2012 *Appl. Phys. Lett.* **101** 023503
[29] Patterson B D *et al* 2010 *New J. Phys.* **12** 035012
[30] Yun K *et al* 2019 *Sci. Rep.* **9** 3300
[31] Bangerter R O 1999 *Phil. Trans. R. Soc. Lond.* A **357** 575–93
[32] Sharkov B Y, Hoffmann D H H, Golubev A A and Zhao Y 2016 *Matter Radiat. Extremes* **1** 28–47
[33] Spiller P and Franchetti G 2006 *Nucl. Instrum. Methods Phys. Res.* A **561** 305–9
[34] Tahir N A *et al* 2005 *Nucl. Instrum. Methods Phys. Res.* A **544** 16–26
[35] Cheng R *et al* 2015 *Matter Radiat. Extremes* **3** 85–93
[36] Friedman A *et al* 2009 *Nucl. Instrum. Meth. Phys. Res.* A **606** 6–10
[37] Tahir N A *et al* 2005 *Contrib. Plasma Phys.* **45** 229–35
[38] Haines M G 2011 *Plasma Phys. Control. Fusion* **53** 093001
[39] Alexandrov V V *et al* 2002 *IEEE Trans. Plasma Sci.* **30** 559–66
[40] Mitchell I H *et al* 1993 *AIP Conf. Proc.* **299** 486–94
[41] Ryutov D D, Derzon M S and Matzen M K 2000 *Rev. Mod. Phys.* **72** 167–223
[42] Spielman R B *et al* 1998 *Phys. Plasmas* **5** 2105–11
[43] Bailey J E *et al* 2002 *Phys. Plasmas* **9** 2186–94

Lightning Source UK Ltd.
Milton Keynes UK
UKHW052105180621
385769UK00005B/153